SpringerBriefs in Electrical and Computer Engineering

Control, Automation and Robotics

For further volumes:
http://www.springer.com/series/10198

Federico Bribiesca Argomedo
Emmanuel Witrant · Christophe Prieur

Safety Factor Profile Control in a Tokamak

Federico Bribiesca Argomedo
Emmanuel Witrant
Christophe Prieur
GIPSA-Lab, Department of Automatic
Control Grenoble Campus
University of Grenoble
Saint-Martin-d'Hères
France

ISSN 2192-6786 ISSN 2192-6794 (electronic)
ISBN 978-3-319-01957-4 ISBN 978-3-319-01958-1 (eBook)
DOI 10.1007/978-3-319-01958-1
Springer Cham Heidelberg New York Dordrecht London

Library of Congress Control Number: 2013949719

Printed on acid-free paper

Springer is part of Springer Science+Business Media (www.springer.com)

Preface

Tokamak reactors pose a myriad of interesting control challenges that have been tackled (with more or less success) over the past few decades. However, until recently, most of the control objectives were formulated in terms of one or a few scalar parameters or quantities (such as the plasma position, shape, total plasma current, etc.). In the last few years, more complex control problems, involving spatially distributed quantities (such as the plasma current density, temperature, magnetic field, etc.) that also evolve in time (and are thus represented by partial differential equations) have begun to appear in the literature. One of such problems is the control of the safety factor profile in a tokamak. This book presents a series of techniques and new methods applied to this control problem.

Although some results presented in this book are based on a series of articles by the authors, a great deal of effort was made to present them as a coherent body of work, as opposed to individual self-contained results, showing the logical progression of the research carried by the authors. Since this book should be accessible to the widest possible audience, we have opted to provide only sketches of proofs whenever possible in order to give to the reader the possibility to understand the main concepts without reducing the readability of the content.

In Chap. 1, we present a brief introduction to the control of thermonuclear fusion and highlight the interest of regulating the safety factor. A detailed explanation of the distributed reference model is provided in Chap. 2, as well as a general statement of the control problem and its main technical difficulties. A finite-dimensional control approach is proposed in Chap. 3, after discretizing the partial differential equations that model the poloidal magnetic flux evolution. The partial fulfillment of the control objectives obtained with this approach motivates the distributed approach that is followed in the rest of the book. Chapter 4 details the main theoretical results necessary for developing a distributed control law that can take into account space and time-variations of the transport parameters, together with some simulation results on a control-oriented simulator. Finally, Chap. 5 presents some extensions required for implementing the proposed scheme on a realistic scenario, along with simulation results on more complex, physics-oriented codes.

The authors wish to acknowledge the work of our co-authors and collaborators, who helped us to set the bases of the results presented in this book. We are thus grateful to Sylvain Brémond, Rémy Nouailletas, and Jean-François Artaud at CEA in Cadarache, to Federico Felici and Olivier Sauter at EPFL, Lausanne, and to Didier Georges at GIPSA-Lab, Grenoble. Also, the useful comments of Hans Zwart, Jacques Blum (as opponents), Jonathan Lister, and Thierry Gallay (as examiners) on the Ph.D. dissertation of the first author have found their way into this book.

San Diego, CA, USA, June 2013
Grenoble, France

Federico Bribiesca Argomedo
Emmanuel Witrant
Christophe Prieur

Editors' Biography

Tamer Başar is with the University of Illinois at Urbana-Champaign, where he holds the academic positions of Swanlund Endowed Chair, Center for Advanced Study Professor of Electrical and Computer Engineering, Research Professor at the Coordinated Science Laboratory, and Research Professor at the Information Trust Institute. He received the B.S.E.E. degree from Robert College, Istanbul, and the M.S., M.Phil, and Ph.D. degrees from Yale University. He has published extensively in systems, control, communications, and dynamic games, and has current research interests that address fundamental issues in these areas along with applications such as formation in adversarial environments, network security, and resilience in cyber-physical systems, and pricing in networks.

In addition to his editorial involvement with these *Briefs*, Başar is also the Editor-in-Chief of *Automatica*, Editor of two Birkhäuser Series on *Systems and Control* and *Static and Dynamic Game Theory*, the Managing Editor of the *Annals of the International Society of Dynamic Games* (ISDG), and Member of Editorial and Advisory Boards of several international journals in control, wireless networks, and applied mathematics. He has received several awards and recognitions over the years, among which are the Medal of Science of Turkey (1993); Bode Lecture Prize (2004) of IEEE CSS; Quazza Medal (2005) of IFAC; Bellman Control Heritage Award (2006) of AACC; and Isaacs Award (2010) of ISDG. He is a member of the US National Academy of Engineering, Fellow of IEEE and IFAC, Council Member of IFAC (2011–2014), a past president of CSS, the founding president of ISDG, and president of AACC (2010–2011).

Antonio Bicchi is Professor of Automatic Control and Robotics at the University of Pisa. He graduated from the University of Bologna in 1988 and was a postdoc scholar at M.I.T. A.I. Lab between 1988 and 1990.

His main research interests are in:

- dynamics, kinematics and control of complex mechanical systems, including robots, autonomous vehicles, and automotive systems;
- haptics and dextrous manipulation; and

- theory and control of nonlinear systems, in particular hybrid (logic/dynamic, symbol/signal) systems.

He has published more than 300 papers on international journals, books, and refereed conferences.

Professor Bicchi currently serves as the Director of the Interdepartmental Research Center "E. Piaggio" of the University of Pisa, and President of the Italian Association or Researchers in Automatic Control. He has served as Editor-in-Chief of the Conference Editorial Board for the IEEE Robotics and Automation Society (RAS), and as Vice President of IEEE RAS, Distinguished Lecturer, and Editor for several scientific journals including the *International Journal of Robotics Research*, the *IEEE Transactions on Robotics and Automation*, and *IEEE RAS Magazine*. He has organized and co-chaired the first World Haptics Conference (2005), and Hybrid Systems: Computation and Control (2007). He is the recipient of several best paper awards at various conferences, and of an Advanced Grant from the European Research Council. Antonio Bicchi has been an IEEE Fellow since 2005.

Miroslav Krstic holds the Daniel L. Alspach chair and is the founding Director of the Cymer Center for Control Systems and Dynamics at University of California, San Diego. He is a recipient of the PECASE, NSF Career, and ONR Young Investigator Awards, as well as the Axelby and Schuck Paper Prizes. Prof. Krstic was the first recipient of the UCSD Research Award in the area of engineering and has held the Russell Severance Springer Distinguished Visiting Professorship at UC Berkeley and the Harold W. Sorenson Distinguished Professorship at UCSD. He is a Fellow of IEEE and IFAC. Prof. Krstic serves as Senior Editor for *Automatica* and *IEEE Transactions on Automatic Control* and as Editor for the Springer series *Communications and Control Engineering*. He has served as Vice President for Technical Activities of the IEEE Control Systems Society. Krstic has co-authored eight books on adaptive, nonlinear, and stochastic control, extremum seeking, control of PDE systems including turbulent flows, and control of delay systems.

Contents

Chapter 1
Introduction

Global warming of the climate system is now unequivocal, mostly due to greenhouse gases (GHGs, increased by 70 % between 1970 and 2004) and aerosols, with a dominant contribution of CO_2 emissions [8]. CO_2 from fossil fuel use amounts to 56.6 % of the anthropogenic GHG emissions. The need for carbon-free energy resources thus appears as a first priority.

Controlled thermonuclear fusion could be a solution to produce sustainable energy. Indeed, harnessing the energy generated by the fusion of Deuterium and Tritium (isotopes of Hydrogen extracted from water and the earth's crust) can be done in a harmless way (no direct radioactive waste and rapid decay of the structure radioactivity). The natural abundance of Deuterium and the possibility of producing Tritium from readily available Lithium mean that easily available fuel reserves for this kind of energy production could amount, in all likelihood, to thousands of years of world energy consumption at current levels [37]. Like all nuclear power sources, the absence of carbon emissions is a key advantage of using nuclear fusion. Furthermore, the inherent safety of the fusion reaction (as opposed to the fission one) and the comparatively easy treatment of radioactive by-products (only structural components that are in close proximity to where the reaction takes place become activated and need to be stored for a few decades before being safely recycled) make this form of energy production extremely attractive.

However enticing the prospect of controlled nuclear fusion may be, achieving and maintaining the fusion reaction is not simple. To fuse two (positively charged) nuclei, the electrostatic force keeping them apart must be overcome (Coulomb barrier reduced by quantum-mechanical tunneling). This is done by taking the fuel to extremely high temperatures (which ionize the fuel atoms, forming a plasma). Once the fuel has enough kinetic energy to overcome this barrier, a question that remains is whether or not a significant amount of nuclei will fuse producing a net energy gain. To achieve this (in a technically feasible way), two main approaches have been explored:

F. Bribiesca Argomedo et al., *Safety Factor Profile Control in a Tokamak*,
SpringerBriefs in Control, Automation and Robotics,
DOI: 10.1007/978-3-319-01958-1_1,

- heating the fuel to obtain a high-density plasma (high fusion rate) and keeping it confined for a short period of time, which is the principle behind inertial confinement fusion, and
- heating the fuel to obtain a low-density plasma (low fusion rate) and keeping it confined for a long period of time, which is the principle behind magnetic confinement fusion.

Tokamaks use a magnetic field to confine a plasma in the shape of a torus. The charged particles follow a helicoidal trajectory according to the field created by controlled magnets, which thus set the position and shape of the plasma. Radio-frequency antennas allow selective action on electrons or ions and modify internal plasma properties such as current and temperature. The plasma is fueled by pellets shot at high speed toward the plasma center and neutral particles are injected to increase the plasma momentum and energy.

The ITER Tokamak [16] (see Fig. 1.1), an international project involving seven members (European Union, Russia, USA, Japan, China, Korea, and India), is planned to start its operation during the next decade. It is foreseen to produce 500 MW out of 50 MW of input power, thus competing with the traditional fission power plants. ITER is currently under construction at Cadarache in southern France. Fusion devices using magnetic confinement of the plasma, such as Tokamaks, can thus be envisaged as a major carbon-free energy resource for the future.

1.1 Challenges in Plasma Physics for Tokamaks

A Tokamak is a toroidal chamber in which magnetic coils generate a very strong magnetic field with both toroidal and poloidal components (see Fig. 1.1). In this chamber, the Tritium-Deuterium plasma circulates so that the fusion reaction can take place (a detailed account of tokamak physics can be found in [37]). Tokamak operation presents several challenging control problems, an overview of which can be found in [1, 30, 35, 36]. Until recently, most of the literature considered the control of one or several scalar parameters of the plasma (for example: shape, position, total current, density). In particular, [1] addresses most of these problems.

When dealing with advanced tokamak scenarios (see for instance [14, 32, 41]) it is desirable to have a finer degree of control on some variables. In particular, full profile control of the current density and pressure may be required. Given the high uncertainty in online profile reconstruction and measurements, as well as in modeling of transport phenomena inside the plasma, controlling these internal profiles is a very challenging task and necessitates robust feedback approaches.

Tokamak control is becoming more and more important for the success of magnetic fusion research and will be crucial for ITER (e.g., see [29, 30] and related tutorials). Feedback control of the main plasma macroscopic parameters, such as plasma position and shape, total current, or density is now reasonably well mastered in the different worldwide Tokamaks. But the control of internal plasma dynamics

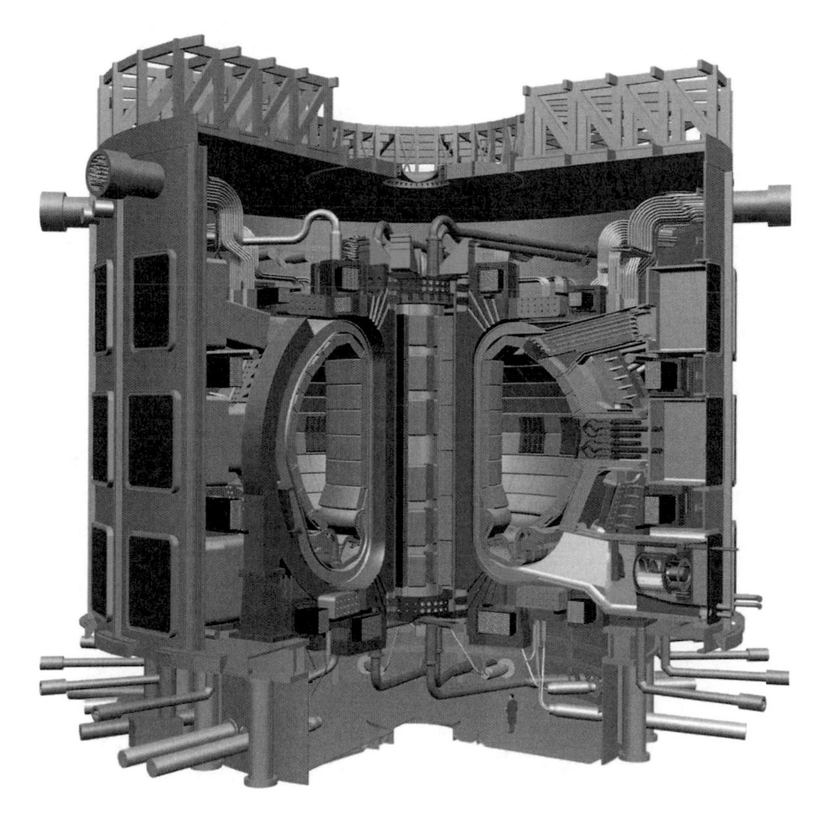

Fig. 1.1 ITER Tokamak (www.iter.org)

and radial profiles is still in its infancy. This control is likely to be crucial for robust stability and to maintain high-efficiency tokamak operation.

According to the plasma physics issues highlighted for ITER, five different objectives can be mentioned:

O.1 *Magneto-hydro-dynamics (MHD) stability*: non-axisymmetric electric currents cause perturbed magnetic fields within (e.g. magnetic islands [34]) or outside (e.g. resistive wall modes [24, 25]) of the plasma, as well as central plasma relaxations (e.g. sawteeth [40]). These instabilities evolve at a fast timescale ($\approx 10^{-6} - 10^{-3}$ s) and need to be adressed in both the poloidal and toroidal directions.

O.2 *Heat confinement*: the fusion reaction efficiency ultimately depends on our capability to raise the central temperature of the plasma to very high values ($\approx 150 \times 10^6$ K, 10 times the central temperature of the sun) while having an edge temperature that can be sustained by the plasma-facing components. Maintaining large temperature gradients is thus essential to achieve an efficient "burn" control while preserving the plasma shell. Confinement predictions are based on empirical scaling laws, non-dimensionally similar studies, or 1-D transport descriptions [17]. A strong limitation for model-based control comes from the difficulty to model the heterogeneous heat transport and internal transport barriers. First approaches on feedback design have been attempted for the burn control using a 0-D approach (e.g. see [5] and references therein).

O.3 *Steady-state operation*: relates to our ability to continue the Tokamak operation indefinitely, e.g. the pulse is terminated by the operator's choice and not due to the plasma behaviour. To this aim, the so-called "safety-factor" (dynamics evolving at the current density diffusion time ≈ 1–100 s) and the pressure profiles provide an indicator on the potential avoidance of MHD instabilities [33]. Based on control-oriented models of the safety-factor profile (such as [20, 39]), lumped [19, 21, 28] and distributed [6, 7] control strategies have been proposed.

O.4 *Control of plasma purity*: an impurity flux is driven by different transport phenomena (e.g., ashes transport, gas puffing at the plasma boundary, and impurity removal) as well as plasma-wall interactions. This problem is related to both preliminary design (e.g., optimal divertor and plasma-facing components) and real-time feedback. This topic and the next one are more specifically associated with extended burn plasmas (to be explored with ITER) and may become first priorities in the near future [31].

O.5 *Exploration of the new physics with a dominant α-particles plasma self heating*: the α-particles (He^{2+}) produced by the fusion reaction are trapped by the magnetic field and transfer their energy to the plasma. They thus provide an extra heat source and induce a local nonlinear feedback. Controlling such an effect would imply to combine anisotropic transport analysis and burn control. A first attempt, focused on maximizing the bootstrap effect on the magnetic flux, is proposed in [12].

While each "physical challenge" is mostly considered as an independent control problem, the automation system will ultimately have to deal with the strong couplings that exist between the different plasma dynamics and the multiple roles of each actuator [18]. For example, the safety factor profile is a key parameter for both the global stability of plasma discharges and an enhanced confinement of the plasma energy (O.1 and O.3), which may reduce the size and cost of future fusion reactors. Other examples include the couplings between the internal variables (e.g., the temperature profile strongly affects the safety factor dynamics, thus coupling objectives O.2 and O.3) and the multiple effects of each actuator (e.g., using the antenna at electron cyclotron radiofrequency mostly affects objectives O.1, O.3, and O.5).

1.2 Control Challenges for Distributed Parameter Systems

The control issues associated with tokamaks involve the spatiotemporal dynamics of transport phenomena (magnetic flux, heat, densities, etc.) in the anisotropic plasma medium. For "steady-state" operation, the state-space variables can be averaged on surfaces of identical magnetic flux (identified with different colors in Fig. 1.2) and the

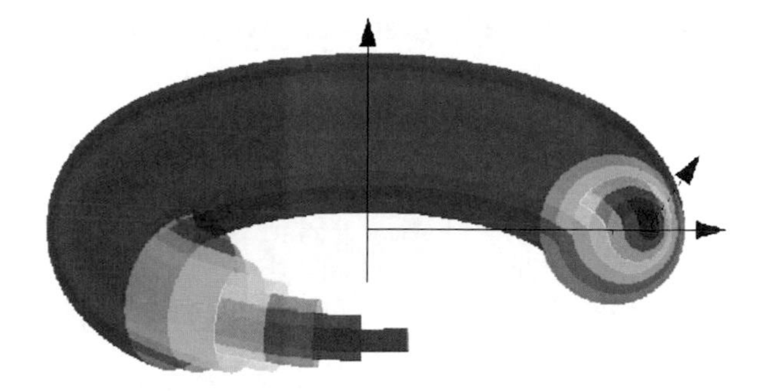

Fig. 1.2 Ideal model of a Tokamak plasma for profiles control, with structured isoflux surfaces and the radial direction

radial profile is regulated in the 1-D space. Both boundary and distributed controls and measurements are available. The physical models typically involve inhomogeneous partial differential equations (PDEs, mostly of parabolic or hyperbolic type) with transport coefficients that differ by several orders of magnitude depending on their location and involve nonlinear couplings between the physical variables. New results are thus sought on:

- **Identification and estimation** possibly with unknown inputs, of transport parameters varying in time and space (Objectives O.1–O.2). Due to the lack of accurate physical models (e.g., for temperature diffusion or internal transport barriers), such results are of prime interest for model-based control and process supervision. Furthermore, the wide ranging of tokamak instrumentation provides an exceptionally rich database for evaluating new estimation strategies in the PDE framework.
- **Stabilization with computation constraints** for high-order linear systems with multiple time-varying delays. Such models can be used to describe convective transport and MHD instabilities based on modal analysis (Objective O.1). The fast timescale of these instabilities prevents the use of full MHD or PDE models and simple feedback laws should be preferred, to control (relatively) large arrays of sensors and actuators (e.g., see [23] for a PID approach on EXTRAP-T2R).
- **Robust PDE control** of 1-D transport equations for the regulation of surfaced-average physical quantities (e.g., safety factor, temperature, and density), which results in a profile control in the radial direction. This relates to several specific problems, such as feedback with timescales separation (Objective O.2), control of linear parameter-varying PDEs (Objective O.3), boundary control (Objective O.4), and nonlinear optimal feedback design for coupled PDEs (Objective O.5).
- **Optimal reference design** to provide scenarios that integrate the multiple coupling constraints for the feedback strategies. These scenarios can be computed offline and use advanced physical models that include the complexity of plasma interconnected dynamics.

1.3 Problem Statement and Background

This book is focused on the feedback control of the *safety factor profile* or *q-profile*. The safety factor corresponds to the number of toroidal versus poloidal rotations done by a field line and is determined by the relation between the two components of the magnetic field. This physical quantity is related to several phenomena in the plasma, in particular, the appearance of MHD instabilities. While the q-profile evolution results from dynamics that are inherently stable, a tight control is necessary to avoid operating conditions at which the MHD instabilities may appear.

Having an adequate safety factor profile is thus particularly important to achieve advanced tokamak operation, providing high confinement and MHD stability. To achieve this, we focus on controlling the poloidal magnetic flux profile (and in particular, its gradient). This is a challenging problem for several reasons [7]:

- the evolution of the magnetic flux is governed by anisotropic diffusion, which is a parabolic equation with spatially distributed rapidly time-varying coefficients that depend on the solution of another partial differential equation related to heat transport;
- the control action is distributed in the spatial domain but nonlinear constraints are imposed on its shape (with only a few engineering parameters available for control, strong restrictions on the admissible shape are imposed);
- nonlinear source terms appear in the evolution equation, (in particular the bootstrap current);
- important uncertainties exist in most measurements, estimations, and models.

The regulation of the poloidal magnetic flux profile is related, by means of the Maxwell equations, to the control of the plasma current profile. The possibility of controlling profile shape parameters has been previously shown in the literature. In [38], the shape of the current profile shape is characterized by the internal inductance of the plasma and the central value of the safety factor, and experimental results are presented. In [4], the control of the width of the lower hybrid power deposition profile is shown and validated with experiments. A discrete real-time control of steady-state safety factor profile is proposed by [15], considering several possible operating modes. Other works consider the distributed nature of the system and use discretized linear models identified around experimental operating point. For example in [19], where a model based on a Galerkin projection was used to control multiple profiles in JET; [22], where a reduced order model is used to control some points of the safety factor profile; [20, 21], where the applicability of these identification and integrated control methods to various Tokamaks is analyzed; and in [26] where a robust controller is constructed based on a POD/Galerkin decomposition and assuming diffusivity profiles with constant shape (varying only modulo a scalar quantity).

The automatic control research community has also made some recent contributions based on simplified control-oriented models that retain the distributed nature of the system. Some examples are [27], where an extremum-seeking open-loop optimal control is developed for the current profile in DIII-D, and [9], where non-linear model-based optimization is used to compute open-loop actuator trajectories for plasma profile control. For closed-loop control examples, in [28] and related works, an infinite-dimensional model, described by partial differential equations (PDEs), is used to construct an optimal controller for the current profile, considering fixed shape profiles for the current deposited by the Radio Frequency (RF) antennas and for the diffusivity coefficients. Other PDE-control approaches, related to Tore Supra, can be mentioned: [11], where sum-of-square polynomials are used to construct a Lyapunov function considering constant diffusivity coefficients; [13], where a sliding-mode controller was designed for the infinite-dimensional system, considering time-invariant diffusivity coefficients; [12], where the nonlinear source term (bootstrap current) is maximized using a Lyapunov-based feedback strategy that employs the sum-of-squares framework.

The derivation of a simplified control-oriented model of the poloidal magnetic flux diffusion is first recalled in Chap. 2. It is based on [39]. This model is composed of a diffusion-like parabolic partial differential equation with time-varying distributed parameters. These types of PDEs (in particular diffusion or diffusion-convection equations) are used to model a wide array of physical phenomena ranging from heat conduction to the distribution of species in biological systems. While the diffusivity coefficients can be assumed to be constant throughout the spatial domain for most applications, spatially distributed coefficients are needed when treating inhomogeneous or anisotropic (direction-dependent) media. Unfortunately, extending existing results from the homogeneous to the inhomogeneous case is not straightforward, particularly when the transport coefficients are time-varying.

1.4 Main Contributions

The goals of this book are:

- the illustration of some control schemes that result from the discretization of the distributed model before designing a control law (lumped-parameter approach) and their inherent limits;
- the use of a simplified infinite-dimensional model for the development of a distributed control law to track the gradient of the magnetic flux profile by means of a Lower Hybrid Current Drive actuator, with specific treatment of time-varying effects and the possible extension to other non-inductive current sources;
- the consideration of time-varying profiles for the diffusivity coefficients in the control design to guarantee stability and robustness of the system with respect to several common sources of errors as well as unmodeled dynamics;
- the inclusion of couplings between the total plasma current control and the magnetic flux profile control;

- a real-time optimization strategy that includes the nonlinear constraints imposed by the current deposit profiles while preserving the theoretical stability and robustness guarantees;
- the validation of the proposed control approach using the METIS code [2] (a module of the CRONOS suite of codes, suitable for closed-loop control simulations [3]) for Tore Supra;
- the addition of profile-reconstruction delays in the control loop;
- the extension of the control scheme to Electron Cyclotron Current Drive (ECCD) actuators and simulation using the RAPTOR code [10] for the TCV experiment (EPFL, Lausanne, Switzerland).

1.5 Outline

This book is organized as follows:

- Chapter 2 presents the main distributed model and its derivation from the magnetohydrodynamics equations. It is used throughout this book along with the physical hypotheses required for the model simplification;
- Chapter 3 presents a control approach based on the spatial discretization of the distributed model presented in Chap. 2. The time-varying character of the diffusivity profiles implies linear matrix inequalities (LMIs) inferred from a polytopic linear parameter-varying (LPV) structure of the model. These LMIs allow computing stabilizing controllers for the extreme variations of the parameters (the vertices of the polytope). Even though this approach takes into account the variation of the diffusivity coefficients, its extension to the control of the gradient of the poloidal flux profile is not straightforward;
- In Chap. 4, a strict Lyapunov function is derived for the open-loop distributed system, which allows constructing strongly constrained control laws that preserve the stability of the system while modifying the input-to-state gains between different disturbances and the gradient of the magnetic flux. Alternative Lyapunov functions are compared to motivate our final choice;
- Chapter 5 includes the important couplings between the Lower Hybrid (LH) power and the total plasma current in the tokamak in the distributed control design method. Advanced simulations using the METIS code illustrate the robustness of the control scheme with respect to modeling errors, input disturbances (represented, for example, by the Ion Cyclotron Resonance Heating (ICRH) actuator) and profile-reconstruction delays. Finally, the flexibility of the proposed scheme to accomodate different actuator models and plasma shapes is illustrated with some simulations using the RAPTOR code [10] for the TCV specifications;
- Chapter 6 contains some concluding remarks and suggests some possible future research lines.

References

1. M. Ariola, A. Pironti, *Magnetic Control of Tokamak Plasmas*. Advances in Industrial Control (Springer, London, 2008)
2. J. F. Artaud. METIS user's guide. CEA/IRFM/PHY/NTT-2008.001, 2008
3. J.F. Artaud et al., The CRONOS suite of codes for integrated Tokamak modelling. Nucl. Fusion **50**, 043001 (2010)
4. O. Barana, D. Mazon, L. Laborde, F. Turco, Feedback control of the lower hybrid power deposition profile on Tore Supra. Plasma Phys. Control. Fusion **49**, 947–967 (2007)
5. M.D. Boyer, E. Schuster. Adaptive nonlinear burn control in tokamak fusion reactors. In *Proceedings of the 2012 American Control Conference*, Montreal, Canada, June 2012, pp. 5043–5048
6. F. Bribiesca Argomedo, C. Prieur, E. Witrant, S. Brémond, A strict control Lyapunov function for a diffusion equation with time-varying distributed coefficients. IEEE Trans. Autom. Contr. **58**(2), 290–303 (2013)
7. F. Bribiesca Argomedo, E. Witrant, C. Prieur, S. Brémond, R. Nouailletas, J.F. Artaud, Lyapunov-based distributed control of the safety-factor profile in a tokamak plasma. Nucl. Fusion **53**(3), 033005 (2013)
8. Core Writing Team, R.K. Pachauri, A. (eds.) Reisinger. Climate change, *Synthesis report* (Contribution of Working Groups I, II and III to the Fourth Assessment Report of the Intergovernmental Panel on Climate Change, 2007) 2007
9. F. Felici, O. Sauter, Non-linear model-based optimization of actuator trajectories for tokamak plasma profile control. Plasma Phys. Contr. Fusion **54**, 025002 (2012)
10. F. Felici, O. Sauter, S. Coda, B.P. Duval, T.P. Goodman, J-M. Moret, J.I. Paley, and the TCV Team, Real-time physics-model-based simulation of the current density profile in tokamak plasmas. Nucl. Fusion **51**, 083052 (2011)
11. A. Gahlawat, M. M. Peet, E. Witrant, Control and verification of the safety-factor profile in tokamaks using sum-of-squares polynomials. In *Proceedings of the 18th IFAC World Congress*. Milan, Italy, August 2011
12. A. Gahlawat, E. Witrant, M.M. Peet, M. Alamir, Bootstrap current optimization in tokamaks using sum-of-squares polynomials, in *Proceedings of the 51st IEEE Conference on Decision and Control* (Maui, Hawaii, 2010), pp. 4359–4365
13. O. Gaye, E. Moulay, S. Brémond, L. Autrique, R. Nouailletas, Y. Orlov, Sliding mode stabilization of the current profile in tokamak plasmas, in *Proceedings of the 50th IEEE Conference on Decision and Control and European Control Conference* (Orlando, FL., 2011), pp. 2638–2643
14. C. Gormezano, High performance tokamak operation regimes. Plasma Phys. Contr. Fusion **41**, B367–B380 (1999)
15. F. Imbeaux et al., Real-time control of the safety factor profile diagnosed by magneto-hydrodynamic activity on the Tore Supra tokamak. Nucl. Fusion **51**, 073033 (2011)
16. ITER Organization. Official ITER site: http://www.iter.org/ Accessed 6 Oct 2010
17. ITER Physics Expert Groups. Chapter 2: Plasma confinement and transport. Nucl. Fusion, **39**(12), 2175–2249 (1999)
18. K. Kurihara, J. B. Lister, D. A. Humphreys, J. R. Ferron, W. Treutterer, F. Sartori, R. Felton, S. Brémond, P. Moreau, JET EFDA contributors. Plasma control systems relevant to iter and fusion power plants. Fusion Eng. Des. **83**, 959–970 (2008)
19. L. Laborde et al., A model-based technique for integrated real-time profile control in the JET tokamak. Plasma Phys. Control. Fusion **47**, 155–183 (2005)
20. D. Moreau, D. Mazon, M.L. Walker, J.R. Ferron, K.H. Burrell, S.M. Flanagan, P. Gohil, R.J. Groebner, A.W. Hyatt, R.J. La Haye, J. Lohr, F. Turco, E. Schuster, Y. Ou, C. Xu, Y. Takase, Y. Sakamoto, S. Ide, T. Suzuki, and ITPA-IOS group members and experts. Plasma models for real-time control of advanced tokamak scenarios. Nucl. Fusion, **51**(6), 063009 (2011)
21. D. Moreau, M.L. Walker, J.R. Ferron, F. Liu, E. Schuster, J.E. Barton, M.D. Boyer, K.H. Burrell, S.M. Flanagan, P. Gohil, R.J. Groebner, C. T. Holcomb, D.A. Humphreys, A.W. Hyatt,

R.D. Johnson, R.J. La Haye, J. Lohr, T.C. Luce, J.M. Park, B.G. Penaflor, W. Shi, F. Turco, W. Wehner, and ITPA-IOS group members and experts. Integrated magnetic and kinetic control of advanced tokamak plasmas on DIII-D based on data-driven models. Nucl. Fusion, **53**, 063020 (2013)
22. P. Moreau et al., Plasma control in Tore Supra. Fusion Sci. Tech. **56**, 1284–1299 (2009)
23. E. Olofsson, E. Witrant, C. Briat, S.-I. Niculescu, and P. Brunsell. Stability analysis and model-based control in extrap-t2r with time-delay compensation. In *47th IEEE Conference on Decision and Control, CDC 2008*, pp. 2044–2049, 2008
24. K. Erik, J. Olofsson, Per R Brunsell, Emmanuel Witrant, and James R Drake. Synthesis and operation of an fft-decoupled fixed-order reversed-field pinch plasma control system based on identification data. Plasma Phys. Contr. Fusion **52**(10), 104005 (2010)
25. K.E.J. Olofsson, Nonaxisymmetric experimental modal analysis and control of resistive wall MHD in RFPs. PhD thesis, Kungliga Tekniska Högskolan, School of Electrical Engineering (EES), Stockholm, Sweden, 2012
26. Y. Ou, C. Xu, E. Schuster, Robust control design for the poloidal magnetic flux profile evolution in the presence of model uncertainties. IEEE Trans. Plasma Sci. **38**(3), 375–382 (2010)
27. Y. Ou, C. Xu, E. Schuster, T.C. Luce, J.R. Ferron, M.L. Walker, D.A. Humphreys, Design and simulation of extremum-seeking open-loop optimal control of current profile in the DIII-D tokamak. Plasma Phys. Control. Fusion **50**, 115001 (2008)
28. Y. Ou, C. Xu, E. Schuster, T.C. Luce, J.R. Ferron, M.L. Walker, D.A. Humphreys, Optimal tracking control of current profile in tokamaks. IEEE Trans. Contr. Syst. Tech. **19**(2), 432–441 (2011)
29. A. Pironti, M. Walker, Control of tokamak plasmas: introduction to a special section. IEEE Contr. Syst. Mag. **25**(5), 24–29 (2005)
30. A. Pironti, M. Walker, Fusion, tokamaks, and plasma control: an introduction and tutorial. IEEE Contr. Syst. Mag. **25**(5), 30–43 (2005)
31. P.H. Rutherford (The Physics Role of ITER. Plasma Physics Laboratory, Princeton University, 1997)
32. T.S. Taylor, Physics of advanced tokamaks. Plasma Phys. Contr. Fusion **39**, B47–B73 (1997)
33. F. Troyon, R. Gruber, H. Saurenmann, S. Semenzato, S. Succi. MHD-limits to plasma confinement. Plasma Phys. Contr. Fusion, **26**(1A), 209 (1984)
34. F. Turco, G. Giruzzi, J.-F. Artaud, O. Barana, V. Basiuk, G. Huysmans, F. Imbeaux, P. Maget, D. Mazon, J.-L. Ségui. Investigation of MHD phenomena on Tore Supra by localised eccd perturbation experiments. IOP Plasma Phys. Contr. Fusion, **50**, 035001 (22pp) (2008)
35. M.L. Walker, E. Schuster, D. Mazon, D. Moreau, Open and emerging control problems in tokamak plasma control, in *Proceedings of the 47th IEEE Conference on Decision and Control* (Cancún, Mexico, 2008), pp. 3125–3132
36. M.L. Walker, D.A. Humphreys, D. Mazon, D. Moreau, M. Okabayashi, T.H. Osborne, E. Schuster, Emerging applications in tokamak plasma control. IEEE Contr. Syst. Mag. **26**(2), 35–63 (2006)
37. J. Wesson, Tokamaks. *International Series of Monographs on Physics 118*, (Oxford University Press, Oxford, 2004)
38. T. Wijnands, D. Van Houtte, G. Martin, X. Litaudon, P. Froissard, Feedback control of the current profile on Tore Supra. Nucl. Fusion **37**, 777–791 (1997)
39. E. Witrant, E. Joffrin, S. Brémond, G. Giruzzi, D. Mazon, O. Barana, P. Moreau, A control-oriented model of the current control profile in tokamak plasma. Plasma Phys. Contr. Fusion **49**, 1075–1105 (2007)
40. G. Witvoet, Feedback control and injection locking of the sawtooth oscillation in fusion plasmas. PhD thesis, Technische Universiteit Eindhoven, Department of Mechanical Engineering, Eindhoven, The Netherlands, 2011
41. R.C. Wolf, Internal transport barriers in tokamak plasmas. Plasma Phys. Contr. Fusion **45**, R1–R91 (2003)

Chapter 2
Mathematical Model of the Safety Factor and Control Problem Formulation

We are interested in controlling the safety factor profile in a tokamak plasma. As the safety factor depends on the ratio of the normalized radius to the poloidal magnetic flux gradient, controlling the gradient of the magnetic flux allows controlling the safety factor profile. In this chapter we present the reference dynamical model [1] for the poloidal magnetic flux profile and its gradient (equivalent to the effective poloidal field magnitude, as defined in [2]), used throughout the following chapters, as well as the control problem formulation. Some of the main difficulties encountered when dealing with this problem are also highlighted.

2.1 Inhomogeneous Transport of the Poloidal Magnetic Flux

The poloidal magnetic flux, denoted $\psi(R, Z)$, is defined as the flux per radian of the magnetic field $B(R, Z)$ through a disc centered on the toroidal axis at height Z, having a radius R, see Fig. 2.1. A simplified one-dimensional model for this poloidal magnetic flux profile is considered. Its dynamics is given by the following equation [3]:

$$\frac{\partial \psi}{\partial t} = \frac{\eta_{\parallel} C_2}{\mu_0 C_3} \frac{\partial^2 \psi}{\partial \rho^2} + \frac{\eta_{\parallel} \rho}{\mu_0 C_3^2} \frac{\partial}{\partial \rho} \left(\frac{C_2 C_3}{\rho} \right) \frac{\partial \psi}{\partial \rho} + \frac{\eta_{\parallel} \frac{\partial \mathscr{V}}{\partial \rho} B_{\phi_0}}{F C_3} j_{ni} \tag{2.1}$$

with the geometry defined by:

$$\rho \doteq \sqrt{\frac{\phi}{\pi B_{\phi_0}}}, \quad C_2(\rho) \doteq \frac{\partial \mathscr{V}}{\partial \rho} \langle \frac{\|\rho\|^2}{R^2} \rangle, \quad C_3(\rho) \doteq \frac{\partial \mathscr{V}}{\partial \rho} \langle \frac{1}{R^2} \rangle$$

where $\langle \cdot \rangle$ represents the average over a flux surface, indexed by the equivalent radius ρ. The remaining coefficients are: ϕ, the toroidal magnetic flux; B_{ϕ_0}, the value of the toroidal magnetic flux at the plasma center; $\eta_{\parallel}$, the parallel resistivity of the plasma; and μ_0, the permeability of free space. j_{ni} is a source term representing the

F. Bribiesca Argomedo et al., *Safety Factor Profile Control in a Tokamak*, SpringerBriefs in Control, Automation and Robotics,
DOI: 10.1007/978-3-319-01958-1_2,

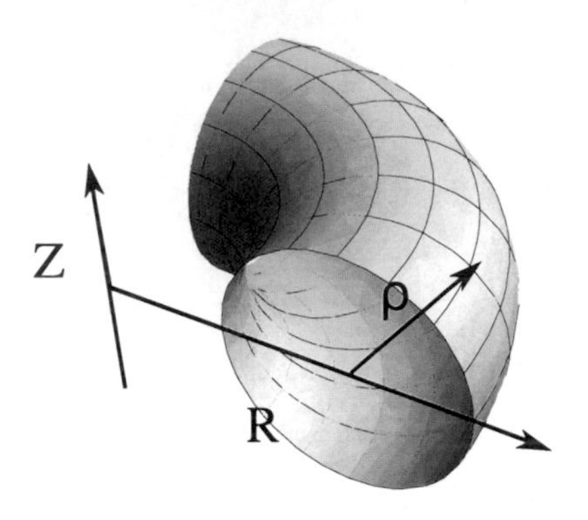

Fig. 2.1 Coordinate system (R, Z) used in this chapter

total effective current produced by non-inductive current sources. F is the diamagnetic function and $\frac{\partial \mathscr{V}}{\partial \rho}$ is the spatial derivative of the plasma volume enclosed by the magnetic surface indexed by ρ. A summary of these variables can be found in Appendix B.

With the following simplifying assumptions:

- $\rho << R_0$ (usually referred to as the cylindrical approximation, where R_0 is the major radius of the plasma);
- the diamagnetic effect caused by poloidal currents can be neglected, the coefficients C_2, C_3 and F simplify to:

$$F \approx R_0 B_{\phi_0}, \quad C_2(\rho) = C_3(\rho) = 4\pi^2 \frac{\rho}{R_0}$$

and the spatial derivative of the enclosed plasma volume becomes:

$$\frac{\partial \mathscr{V}}{\partial \rho} = 4\pi^2 \rho R_0$$

Introducing the normalized variable $r \doteq \rho/a$, a being the minor radius of the last closed magnetic surface, we obtain the simplified model [1, 4]:

$$\frac{\partial \psi}{\partial t}(r,t) = \frac{\eta_{\|}(r,t)}{\mu_0 a^2} \left(\frac{\partial^2 \psi}{\partial r^2} + \frac{1}{r} \frac{\partial \psi}{\partial r} \right) + \eta_{\|}(r,t) R_0 j_{ni}(r,t) \tag{2.2}$$

with the boundary condition at the plasma center:

$$\frac{\partial \psi}{\partial r}(0,t) = 0 \tag{2.3}$$

one of the two boundary conditions at the plasma edge:

$$\frac{\partial \psi}{\partial r}(1, t) = -\frac{R_0 \mu_0 I_p(t)}{2\pi} \quad or \quad \frac{\partial \psi}{\partial t}(1, t) = V_{loop}(t) \tag{2.4}$$

(where I_p is the total plasma current and V_{loop} is the toroidal loop voltage) and with the initial condition:

$$\psi(r, t_0) = \psi_0(r)$$

Remark 2.1 The validity of this model (derived for Tore Supra) can be extended to other tokamaks by changing the definition of the values C_2, C_3, Γ and $\frac{\partial \mathcal{V}}{\partial \rho}$.

2.2 Periferal Components Influencing the Poloidal Magnetic Flux

The dynamics (2.2) depend on the plasma resistivity (diffusion coefficient), the inductive current generated by the poloidal coils (boundary control input) and the non-inductive currents (distributed control input and nonlinearity), which can be described as follows.

2.2.1 *Resistivity and Temperature Influence*

The diffusion term in the magnetic flux dynamics is provided by the plasma resistivity $\eta_\parallel$, which introduces a coupling with the temperature (main influence) and density profiles. This parameter is obtained from the neoclassical conductivity proposed in [5] (approximate analytic approach) using the electron thermal velocity and Braginskii time, computed from the temperature and density profiles as in [6].

The temperature dynamics are typically determined by a resistive-diffusion equation [2], where the diffusion coefficient depends nonlinearly on the safety factor profile [7]. A quasi-1D model was proposed in [8] to model the normalized temperature profiles as scaling laws determined by the global (0-D) plasma parameters and to constrain the temperature dynamics by the global energy conservation. This grey-box model was shown to provide a sufficient accuracy for the magnetic flux prediction in [1].

The resistive-diffusion time is much faster (more than 10 times) than the current density diffusion time, which motivated lumped control approaches based on the separation of the timescales and using a linear time-invariant model [9, 10]. In our case, we consider that $\eta_\parallel$ varies in time and space, to avoid the strong dependency on the operating point, but we do not address specifically the problem of a coupling with the temperature dynamics (thus considering only the linear time-varying contribution of the resistivity).

As an example, the resistivity calculated with measured temperature profiles is depicted in Fig. 2.2. This plasma shot is characterized by power modulations of the

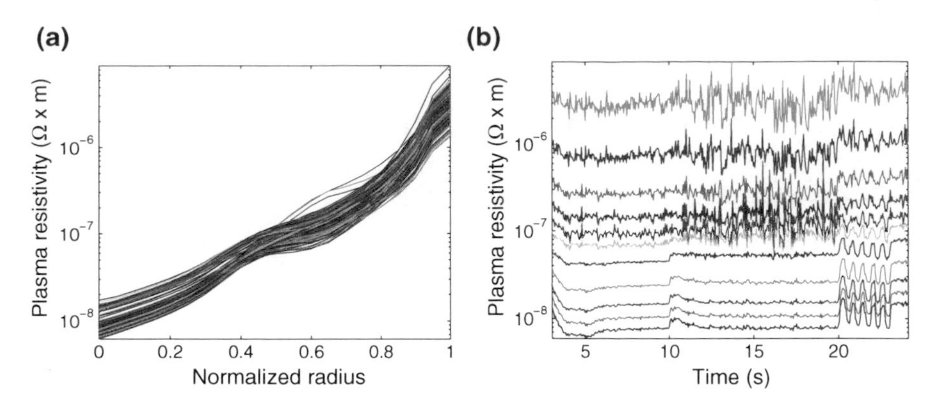

Fig. 2.2 **a** Spatial distribution for different time instants. **b** Time evolution at different locations. Plasma resistivity profiles computed for Tore Supra shot 35952 (modulations on the LH and ICRH antennas)

lower hybrid (LH) and ion cyclotron radio heating (ICRH) antennas. Note the difference of three orders of magnitude between the plasma center and its edge on Fig. 2.2a, and the modulated and noisy time-evolution on Fig. 2.2b. The crewels observed after 20 s result from LH modulations (step inputs) at relatively low power and illustrate an input-to-state coupling effect, as LH is our main input on the magnetic flux.

2.2.2 Inductive Current Sources

The magnetic flux at the boundary $\frac{\partial}{\partial t}\psi(1, t)$ is set by the poloidal coils surrounding the plasma and the central solenoid, and constitutes the inductive current input. This can be described by the classical transformer model where the coils generate the primary circuit while the plasma is the secondary, modelled as a single filament. The dynamics of the coils current $I_c(t)$ and of the plasma current $I_p(t)$ are coupled as [11]:

$$\begin{bmatrix} L_c & M \\ M & L_p \end{bmatrix} \frac{d}{dt} \begin{bmatrix} I_c \\ I_p \end{bmatrix} = - \begin{bmatrix} R_c & 0 \\ 0 & R_p \end{bmatrix} \begin{bmatrix} I_c \\ I_p \end{bmatrix} + \begin{bmatrix} V_c \\ R_p I_{NI} \end{bmatrix}$$

where R_c and L_c are the coils resistance and internal inductance, R_p and L_p are the plasma resistance and inductance, M is the matrix of mutual inductances, V_c is the input voltage applied to the coils and I_{NI} is the current generated by the non-inductive sources. Note that the values of R_c and L_c are given from the coil properties while M is obtained from an equilibrium code (i.e. CEDRES++ [12]). Considering the effects of the plasma current and inductance variations, the loop voltage V_{loop} is obtained from [13] with:

$$V_{loop}(t) = -\frac{1}{I_p}\frac{d}{dt}\left[\frac{L_p I_p^2}{2}\right] + M\frac{dI_c}{dt}$$

In practice, a local control law is set on the poloidal coils to adjust the value of V_c according to a desired value of V_{loop}, which can be measured with a Rogowski coil. If the reference is set on the plasma current I_p instead, then V_c is such that the coils provide the current necessary for completing the non inductive sources to obtain I_p.

2.2.3 Non-inductive Current: Sources and Nonlinearity

The non-inductive current j_{ni} in the magnetic flux dynamics (2.2) is composed of two types of sources: the controlled inputs and the bootstrap effect.

The controlled inputs are the current drive (CD) effects associated with neutral beam injection (NBI), traveling waves in the lower hybrid (LH) frequency range (0.8–8 GHz, the most effective scheme) and electron cyclotron (EC) waves. The precise physical modeling of the CD effects necessitates a complex analysis of the coupling between waves and particles, which cannot be used for real-time control purposes. We consider instead some semi-empirical models for the Lower Hybrid Current Drive (LHCD) and Electron Cyclotron Current Drive (ECCD) antennas (NBI is not explicitly included in our control schemes but the general strategy would remain the same), where the current deposit shape is constrained to fit a gaussian bell [1]. The gaussian shape is identified from experimental data (LHCD) or obtained from model simplifications (ECCD), while the amplitude of the deposit comes from CD efficiency computations involving the density, temperature and total current of the plasma.

For example the shape of LHCD deposit can be adequately approximated by a gaussian curve with parameters μ, σ and A_{lh} (which depend on the engineering parameters P_{lh} and $N_{||}$ and on the operating point) as:

$$j_{lh}(r,t) = A_{lh}(t)e^{-(r-\mu(t))^2/(2\sigma^2(t))}, \forall (r,t) \in [0,1] \times [0,T] \qquad (2.5)$$

Scaling laws for the shape parameters can be built based on suprathermal electron emission, measured via hard X-ray measurements, see for instance [14] and [15]. The total current driven by the LH antenna is then calculated using scaling laws such as those presented in [16]. It should be noted that the methods presented in this book can easily be extended to other current deposit shapes (either for use in other tokamaks or to change the non-inductive current drives used).

While the impact of I_p on the deposit amplitude would induce a nonlinearity (product between the state and the control input), we neglect this effect by considering an extra loop on the radio-frequency antennas that sets the engineering inputs according to a desired profile. Such strategy is motivated by the fact that the antennas react much faster than the plasma and can thus generate a desired profile almost instantly.

The second non-inductive source of current is due to the bootstrap effect, induced by particles trapped in a banana orbit. Expressing the physical model derived by [17] in cylindrical coordinates and in terms of the magnetic flux, the bootstrap current is obtained as:

$$j_{bs}(r,t) = \frac{p_e R_0}{\partial\psi/\partial r}\left\{A_1\left[\frac{1}{p_e}\frac{\partial p_e}{\partial r} + \frac{p_i}{p_e}\left(\frac{1}{p_i}\frac{\partial p_i}{\partial r} - \alpha_i\frac{1}{T_i}\frac{\partial T_i}{\partial r}\right)\right] - A_2\frac{1}{T_e}\frac{\partial T_e}{\partial r}\right\}$$

where $p_{e/i}$ is the electron and ion pressure, $T_{e/i}$ is the electron and ion temperature, α_i depends on the ratio of trapped to circulating particles x_t and $A_{1/2}(r,t)$ depend on x_t and on the effective value of the plasma charge. Maximizing the bootstrap effect, as a "free" source of non-inductive current, is one of the prime objectives for large tokamaks such as ITER, which motivated the bootstrap current maximization strategy proposed in [18]. As the control approaches discussed in this book are focused on linear time-varying strategies (on the lumped and PDE models), we will consider small deviations from an equilibrium bootstrap distribution (given by the reference magnetic flux distribution) and ensure the robustness with respect to these deviations rather than addressing the nonlinearity directly.

To summarize, j_{ni} is considered as the sum of three components:

$$j_{ni} = j_{lh} + j_{eccd} + j_{bs}$$

2.3 Control Problem Formulation

Based on the previous description of the system dynamics and periferal components obtained from a physical analysis of the tokamak plasma, this section discusses the appropriate change of variables to formulate the control problem. We also describe the control objectives and challenges for an efficient regulation of the safety-factor profile, which will be answered in the following chapters.

2.3.1 *Equilibrium and Regulated Variation*

We define $\eta \doteq \eta_{\|}/\mu_0 a^2$ and $j \doteq \mu_0 a^2 R_0 j_{ni}$ to simplify the notations. An equilibrium $\overline{\psi}$, if it exists, is defined as a stationary solution of:

$$0 = \left[\frac{\eta}{r}\left[r\overline{\psi}_r\right]_r\right]_r + \left[\eta\bar{j}\right]_r, \forall r \in (0,1) \tag{2.6}$$

with the boundary conditions:

$$\overline{\psi}_r(0) = 0$$
$$\overline{\psi}_r(1) = -\frac{R_0 \mu_0 \overline{I}_p}{2\pi} \tag{2.7}$$

for a given couple $(\overline{j}, \overline{I}_p)$, where, to simplify the notation, for any function ξ depending on the independent variables r and t, ξ_r and ξ_t are used to denote $\frac{\partial \xi}{\partial r}$ and $\frac{\partial \xi}{\partial t}$, respectively.

Remark 2.2 When seeking an equilibrium by solving (2.6)–(2.7) two cases have to be considered:

(i) there is no drift in $\overline{\psi}$ (equivalent to $V_{loop} = 0$ at all times using the alternative boundary condition $\psi_t(1, t) = V_{loop}(t)$ in (2.4)) and therefore the solution of (2.6)–(2.7) verifies:
$$\frac{\eta}{r} \left[r\overline{\psi}_r \right]_r + \eta \overline{j} = 0 \tag{2.8}$$

In this case, $\overline{\psi}$ (and its spatial derivative) is independent on the value of η and therefore the stationary solution exists (i.e. there is an equilibrium of the time-varying system) regardless of the variations in η. This is the case we directly address in this book.

(ii) there is a radially constant drift in $\overline{\psi}$ (equivalent to $V_{loop} \neq 0$ for some times when using the alternative boundary condition) and therefore the solution of (2.6)–(2.7) verifies, for some $c(t)$:

$$\frac{\eta(r,t)}{r} \left[r\overline{\psi}_r(r,t) \right]_r + \eta(r,t)\overline{j}(r) = c(t) \tag{2.9}$$

In this case, $\overline{\psi}_r$ does not correspond to an equilibrium since it will be a function of time and space (in particular, it will be a function of $\eta(r,t)$ and $c(t)$), we will call the corresponding $\overline{\psi}(r,t)$ a pseudo-equilibrium of the system. It can be shown to verify:

$$\overline{\psi}_r(r,t) = \frac{1}{r} \int_0^r \left(\frac{\rho}{\eta(\rho,t)} c(t) - \rho \overline{j}(\rho) \right) d\rho \tag{2.10}$$

with time-derivative:

$$\overline{\psi}_{rt}(r,t) = \frac{1}{r} \int_0^r \left(\frac{\rho}{\eta(\rho,t)} \dot{c}(t) - \frac{\rho \dot{\eta}(\rho,t)}{\eta^2(\rho,t)} c(t) \right) d\rho \tag{2.11}$$

This case is not extensively addressed in this book but the results presented in Chaps. 4 and 5 will not be severely affected as long as $\dot{c}(t)$ and $\dot{\eta}(r,t)$ are bounded in a suitable way. Since a pseudo-equilibrium will exist at each time, the robustness result presented in Theorem 4.2 can be applied, rewriting the evolution of the system around this pseudo-equilibrium (instead of an actual equilibrium) and considering $w = -\overline{\psi}_{rt}(r, t)$ (the time-varying nature of the pseudo-equilibrium acts as a state-disturbance for the system).

Around the equilibrium (assumed to exist as per the previous remark) and neglecting the nonlinear dependence of the bootstrap current on the state, the dynamics of the system is given by:

$$\tilde{\psi}_t = \frac{\eta}{r}\left[r\tilde{\psi}_r\right]_r + \eta\tilde{j}, \ \forall(r,t) \in (0,1) \times (0,T) \tag{2.12}$$

with boundary conditions:

$$\begin{aligned} \tilde{\psi}_r(0,t) &= 0 \\ \tilde{\psi}_r(1,t) &= -\frac{R_0\mu_0\tilde{I}_p(t)}{2\pi} \end{aligned} \tag{2.13}$$

and initial condition:

$$\tilde{\psi}(r,0) = \tilde{\psi}_0(r) \tag{2.14}$$

where the dependence of $\tilde{\psi} \doteq \psi - \overline{\psi}, \tilde{j} \doteq j - \overline{j}$ and η on (r, t) is implicit; $\tilde{I}_p \doteq I_p - \overline{I}_p$ and $0 < T \leq +\infty$ is the time horizon.

As the safety factor profile depends on the magnetic flux gradient, our focus is on the evolution of $z \doteq \partial\tilde{\psi}/\partial r$ (equivalent to deviations of the effective poloidal field magnitude around an equilibrium), with input $u \doteq \tilde{j}$, defined as:

$$z_t(r,t) = \left[\frac{\eta(r,t)}{r}\left[rz(r,t)\right]_r\right]_r + \left[\eta(r,t)u(r,t)\right]_r, \ \forall(r,t) \in (0,1) \times (0,T) \tag{2.15}$$

with Dirichlet boundary conditions:

$$\begin{aligned} z(0,t) &= 0 \\ z(1,t) &= -\frac{R_0\mu_0\tilde{I}_p(t)}{2\pi} \end{aligned} \tag{2.16}$$

and initial condition:

$$z(r,0) = z_0(r) \tag{2.17}$$

where $z_0 \doteq \left[\tilde{\psi}_0\right]_r$.

Following [19], the following properties are assumed to hold in (2.15):

- $\mathbf{P_1}$: $K \geq \eta(r,t) \geq k > 0$ for all $(r,t) \in [0,1] \times [0,T)$ and some positive constants k and K.

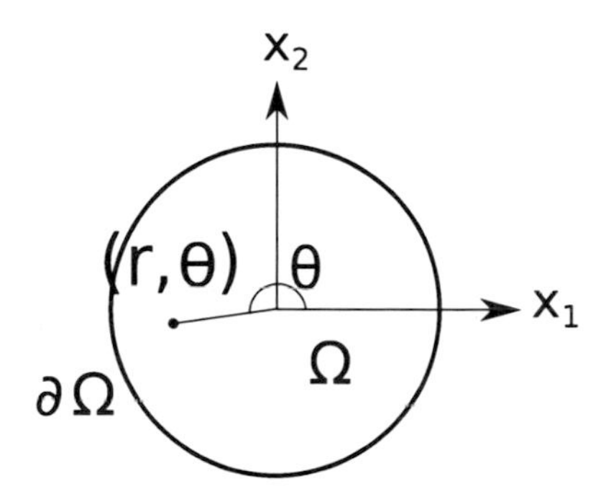

Fig. 2.3 Cartesian coordinates (x_1, x_2), and polar coordinates (r, θ) used in this book

- **P_2**: The two-dimensional Cartesian representations of η and u are in[1] $C^{1+\alpha_c,\alpha_c/2}(\overline{\Omega} \times [0, T])$, $0 < \alpha_c < 1$, where $\Omega \doteq \{(x_1, x_2) \in \mathbb{R}^2 \mid x_1^2 + x_2^2 < 1\}$ as shown in Fig. 2.3.
- **P_3**: $\tilde{I}_p$ is in $C^{(1+\alpha_c)/2}([0, T])$.

For completeness purposes, the existence and uniqueness of sufficiently regular solutions (as needed for the Lyapunov analysis and feedback design purposes in the next chapters) of the evolution equation is stated, assuming that the properties P_1–P_3 are verified.

Theorem 2.1 *If Properties P_1–P_3 hold then, for every $z_0 : [0, 1] \rightarrow \mathbb{R}$ in $C^{2+\alpha_c}([0, 1])$ $(0 < \alpha_c < 1)$ such that $z_0(0) = 0$ and $z_0(1) = -R_0\mu_0\tilde{I}_p(0)/2\pi$, the evolution equations* (2.15)–(2.17) *have a unique solution $z \in C^{1+\alpha_c,1+\alpha_c/2}([0, 1] \times [0, T]) \cap C^{2+\alpha_c,1+\alpha_c/2}([0, 1] \times [0, T])$.*

The proof of this result is given in [19] and mainly follows from [20].

2.3.2 Interest of Choosing ψ as the Regulated Variable

A natural question that may arise at this point is why studying the evolution of the poloidal magnetic flux profile instead of studying directly the safety factor profile. Considering that the safety factor profile is related to the magnetic flux profile as:

$$q(r, t) = -\frac{B_{\phi 0} a^2 r}{\psi_r(r, t)} \tag{2.18}$$

the evolution of the safety factor profile is then given by:

---

[1] Here $C^{\alpha_c,\beta_c}(\overline{\Omega} \times [0, T])$ denotes the space of functions which are α_c-Hölder continuous in $\overline{\Omega}$, β_c-Hölder continuous in $[0, T]$. P_2 can be strengthened by assuming that η and u are in $C^{2,1}(\overline{\Omega} \times [0, T])$ which is the case for the physical application.

$$q_t(r,t) = \frac{B_{\phi_0} a^2 r}{\psi_r^2(r,t)} \psi_{rt}(r,t) = \frac{q^2(r,t)}{B_{\phi_0} a^2 r} \psi_{rt}(r,t)$$

and, from (2.2):

$$q_t(r,t) = -\frac{q^2(r,t)}{r} \left[\frac{\eta(r,t)}{r} \left[\frac{r^2}{q(r,t)} \right]_r \right]_r + \frac{q^2(r,t)}{B_{\phi_0} a^2 r} [\eta(r,t) u(r,t)]_r \tag{2.19}$$

or, in a more general form:

$$\begin{aligned} q_t(\rho,t) = & -\frac{q^2(\rho,t)}{\mu_0 \rho} \left[\frac{\eta_{\parallel}(\rho,t)\rho}{C_3^2(\rho)} \left[\frac{C_2(\rho) C_3(\rho)}{q(\rho,t)} \right]_\rho \right]_\rho \\ & + \frac{q^2(\rho,t)}{\rho} \left[\frac{\eta_{\parallel}(\rho,t) \frac{\partial \mathscr{V}}{\partial \rho}}{F C_3(\rho)} j_{ni}(\rho,t) \right]_\rho \end{aligned}$$

which can be obtained from (2.1) and the relation:

$$q(\rho,t) = -\frac{B_{\phi_0} \rho}{\psi_\rho(\rho,t)}$$

Equation (2.19) is nonlinear in q (making it difficult to extend results obtained around one equilibrium to other equilibria). This can be solved by working instead with the so-called rotational transform (denoted ι in [6], which is the inverse of the safety factor). Nevertheless, the boundary condition in the z variable (i.e. the total plasma current) can be directly (and precisely) measured using either a continuous Rogowski coil or several discrete magnetic coils around the plasma (see [6]). Therefore, in this book, we have chosen to control the safety factor profile by controlling the z variable.

2.3.3 Control Challenges

Controlling the safety factor profile q in a tokamak is done by controlling the poloidal magnetic flux profile ψ. In particular, the desired properties of the controller are:

- to guarantee the exponential stability, in a given topology, of the solutions of equation (2.15) to zero (or "close enough") by closing the loop with a controlled input $u(\cdot,t)$;
- to be able to adjust (in particular, to accelerate) the rate of convergence of the system using the controlled input;
- to be able to determine the impact of a large class of errors motivated by the physical system and to propose a robust feedback design strategy. Actuation errors,

estimation/measurement errors, state disturbances and boundary condition errors should be considered specifically.

The problem under consideration poses several challenges that have to be addressed, some of which are:

- different orders of magnitude of the transport coefficients depending on the radial position that are also time-varying;
- strong nonlinear shape constraints imposed on the actuators and saturations on the available parameters;
- robustness of any proposed control scheme with respect to numerical problems (in particular given the difference in magnitude of the transport coefficients) and disturbances;
- coupling between the control applied to the infinite-dimensional system and the boundary condition;
- the control algorithms must be implementable in real-time (restrictions on the computational cost).

References

1. E. Witrant, E. Joffrin, S. Brémond, G. Giruzzi, D. Mazon, O. Barana, P. Moreau, A control-oriented model of the current control profile in tokamak plasma. Plasma Phys. Control Fusion **49**, 1075–1105 (2007)
2. F.L. Hinton, R.D. Hazeltine, Theory of plasma transport in toroidal confinement systems. Rev. Mod. Phys. **48**(2), 239–308 (April 1976)
3. J. Blum, *Numerical Simulation and Optimal Control in Plasma Physics. Wiley/Gauthier-Villars Series in Modern Applied Mathematics*. (Gauthier-Villars, Wiley, New York, 1989)
4. J.F. Artaud et al., The CRONOS suite of codes for integrated tokamak modelling. Nucl. Fusion **50**, 043001 (2010)
5. S.P. Hirshman, R.J. Hawryluk, B. Birge, Neoclassical conductivity of a tokamak plasma. Nucl. Fusion **17**(3), 611–614 (1977)
6. J. Wesson. *Tokamaks. International Series of Monographs on Physics, 118*, 3rd edn. (Oxford University Press, USA, 2004)
7. M. Erba, A. Cherubini, V.V. Parail, E. Springmann, A. Taroni, Development of a non-local model for tokamak heat transport in L-mode, H-mode and transient regimes. Plasma Phys. Controlled Fusion **39**(2), 261 (1997)
8. E. Witrant, S. Brémond, Shape identification for distributed parameter systems and temperature profiles in tokamaks. in *Decision and Control and European Control Conference (CDC-ECC), 2011 50th IEEE Conference* pp. 2626–2631, 2011
9. L. Laborde et al., A model-based technique for integrated real-time profile control in the JET tokamak. Plasma Phys. Control Fusion **47**, 155–183 (2005)
10. D. Moreau, M.L. Walker, J.R. Ferron, F. Liu, E. Schuster, J.E. Barton, M.D. Boyer, K.H. Burrell, S.M. Flanagan, P. Gohil, R.J. Groebner, C. T. Holcomb, D.A. Humphreys, A.W. Hyatt, R.D. Johnson, R.J. La Haye, J. Lohr, T.C. Luce, J.M. Park, B.G. Penaflor, W. Shi, F. Turco, W. Wehner, and ITPA-IOS group members and experts, Integrated magnetic and kinetic control of advanced tokamak plasmas on DIII-D based on data-driven models. Nucl. Fusion 53:063020, 2013
11. F. Kazarian-Vibert et al., Full steady-state operation in Tore Supra. Plasma Phys. Control Fusion **38**, 2113–2131 (1996)

12. P. Hertout, C. Boulbe, E. Nardon, J. Blum, S. Brémond, J. Bucalossi, B. Faugeras, V. Grandgirard, P. Moreau, The CEDRES++ equilibrium code and its application to ITER, JT-60SA and Tore Supra. Fusion Eng. Des. **86**(6–8), 1045–1048 (2011)
13. N.J. Fisch, Theory of current drive in plasmas. Rev. Mod. Phys. **59**(1), 175–234 (January 1987)
14. F. Imbeaux, Etude de la propagation et de l'absorption de l'onde hybride dans un plasma de tokamak par tomographie X haute énergie. Ph.D. thesis, Université Paris XI, Orsay, France, 1999
15. O. Barana, D. Mazon, G. Caulier, D. Garnier, M. Jouve, L. Laborde, Y. Peyson, Real-time determination of supratermal electron local emission profile from hard x-ray measurements on Tore Supra. IEEE Trans. Nucl. Sci. **53**, 1051–1055 (2006)
16. M. Goniche et al., Lower hybrid current drive efficiency on Tore Supra and JET. 16th Topical Conference on Radio Frequency Power in Plasmas. Park City, USA, 2005
17. S.P. Hirshman, Finite-aspect-ratio effects on the bootstrap current in tokamaks. Phys. Fluids **31**, 3150–3152 (1998)
18. A. Gahlawat, E. Witrant, M.M. Peet, M. Alamir, Bootstrap current optimization in tokamaks using sum-of-squares polynomials, *Proceedings of the 51st IEEE Conference on Decision and Control* (Maui, Hawaii, 2012), pp. 4359–4365
19. F. Bribiesca Argomedo, C. Prieur, E. Witrant, S. Brémond, A strict control Lyapunov function for a diffusion equation with time-varying distributed coefficients. IEEE Trans. Autom. Control **58**(2), 290–303 (2013)
20. A. Lunardi, *Analytic Semigroups and Optimal Regularity in Parabolic Problems, volume 16 of Progress in nonlinear differential equations and their applications*. (Birkhäuser, Basel, 1995)

Chapter 3
A Polytopic LPV Approach for Finite-Dimensional Control

In this chapter, we consider the problem of controlling a discretized version of model (2.2). For control purposes, this model was spatially discretized (in N+2 points) using the (finite-differences) midpoint rule to approximate the operators $\frac{\partial^2}{\partial r^2}$ and $\frac{1}{r}\frac{\partial}{\partial r}$. Details of the process used for the discretization and relevant implementation details can be found in [11]. The calculations are made to allow for a non-uniform spatial step distribution. For simulation purposes, the model was discretized in both space and time following the procedure detailed in [11]. As a first step towards controlling the safety factor profile in the Tokamak, we focus in this chapter on the control of the magnetic flux profile ψ.

After spatially discretizing model (2.2), a Polytopic Linear Parameter-Varying (Polytopic LPV) controller was designed to take into account the transient behavior of the diffusion coefficients in the feedback design. After a change of variables, the time dependence of the (input) matrix B is removed, allowing us to formulate a stabilizing control law for the system (extended with an output integrator) in the form of a convex (linear) combination of suitable controllers calculated for the extrema of the dynamic variations (the vertices of a set of either uncertain or time-varying parameters). This chapter is based on [3]. An extensive literature is available on linear parameter-varying systems (LPV), linear matrix inequalities (LMIs) and gain scheduling, see for example [5, 6, 8, 10].

Discretizing the PDE (2.2) and solving for the boundary points $r = 0$ and $r = 1$, the dynamics of the remaining states is given by:

$$\dot{\psi} = A(t)\psi + B(t)j_{ni} + W(t) \tag{3.1}$$

here $A(t)$ is an $N \times N$ matrix representing the discretized differential operators multiplied by the coefficient $\eta_{\parallel}/(\mu_0 a^2)$. $B(t)$ is an $N \times N$ matrix representing the coefficient $\eta_{\parallel} R_0$. $W(t)$ is an $N \times 1$ column vector representing boundary conditions of the system and possibly other disturbances. For simplicity of notation, we will refer to $\psi(r_i, \cdot)$ simply as $\psi_i(t)$ in this chapter.

F. Bribiesca Argomedo et al., *Safety Factor Profile Control in a Tokamak*,
SpringerBriefs in Control, Automation and Robotics,
DOI: 10.1007/978-3-319-01958-1_3, © The Author(s) 2014

This section is based on results presented in [3]. Our aim is to develop a suitable control law for the regulation of the steady-state magnetic flux profile that allows for a closed-loop stability analysis. In particular, it is based on a polytopic approach similar to the one described in [2]. The tokamak actuation is restricted to the use of the non inductive Lower Hybrid Current Drive (LHCD), which acts as a current and heat source on the plasma (altering both the current source term j_{ni} and the resistivity profile $\eta_{\|}$).

3.1 LPV Model

Since the time-varying coefficient multiplying the source term is a common factor of the diffusion-like operator, Eq. (3.1) can be factorized as follows:

$$\dot{\psi}(t) = M(t)\left(A_{ct}\psi(t) + B_{ct}j_{ni}(t)\right) + W(t) \tag{3.2}$$

where $M(t)$ is an $N \times N$ diagonal matrix (with only positive elements in the diagonal, representing the resistivity coefficient). It should be noted that A_{ct} and B_{ct} are now constant matrices.

Let us define the variation of the state around an equilibrium $(\overline{\psi}, \overline{j}_{ni}, \overline{W})$ (chosen as operating point for the system) as:

$$\begin{aligned} \widetilde{\psi} &\doteq \psi - \overline{\psi} \\ \widetilde{j}_{ni} &\doteq j_{ni} - \overline{j}_{ni} \\ \widetilde{W} &\doteq W - \overline{W} \end{aligned}$$

The values for the equilibrium profiles can be obtained from available experimental data (obtained by using data from different diagnostic systems in the tokamak to reconstruct internal states) or by numerically simulating the evolution of the system.

Assuming that the plasma current is almost constant during steady-state operation and considering the variations of the bootstrap current around the equilibrium as disturbances, the term $\widetilde{W}$ can be neglected for small excursions around the operating point. Furthermore j_{ni} is assummed to be composed of only the LHCD current deposit j_{lh}, which can be approximated as in (2.5).

Linearizing (2.5) with respect to a variation of the parameters around the equilibrium parameters $\overline{u}_p \doteq [\overline{\mu}_{lh}, \overline{\sigma}_{lh}, \overline{A}_{lh}]^T$ (assumed to result in $\overline{j}_{lh}$), and defining the variation of these parameters as $\widetilde{u}_p \doteq u_p - \overline{u}_p$, we obtain the following linearized representation of the system dynamics:

$$\dot{\widetilde{\psi}}(t) = M(t)\left(A_{ct}\widetilde{\psi}(t) + B_{ct}\nabla_{u_p} j_{lh}\,|_{u_p=\overline{u}_p}\,\widetilde{u}_p(t)\right) \tag{3.3}$$

We define a change of variables $\zeta \doteq M^{-1}(t)\widetilde{\psi}$ such that the new input matrix in the evolution equation $B_{lin} \doteq B_{ct}\nabla_{u_p} j_{lh}\,|_{u_p=\overline{u}_p}$ is not time-varying. The evolution of the state ζ is thus given as:

$$\dot{\zeta}(t) = \left(A_{ct}M(t) - M^{-1}(t)\dot{M}(t)\right)\zeta(t) + B_{lin}\widetilde{u}_p(t) \tag{3.4}$$

As $M(t)$ is positive definite, $M^{-1}(t)$ always exists, is positive definite and bounded (in fact, $M(t)$ being a diagonal matrix for all times, the inversion is trivial). Assuming that $M(t)$ is continuously differentiable with respect to time and with bounded derivative, the new dynamic matrix $A_\zeta(t) \doteq A_{ct}M(t) - M^{-1}(t)\dot{M}(t)$ is bounded at all times. Choosing a basis $\mathscr{A} = \{A_{\zeta_0}, A_{\zeta_1}, \ldots, A_{\zeta_{n_p}}\}$, subset of $\mathbb{R}^{N\times N}$, to represent A, we write:

$$A_\zeta(t) = A_{\zeta_0} + \sum_{i=1}^{n_p} \lambda_i(t)A_{\zeta_i} \tag{3.5}$$

with $n_p \leq 2N$, $\lambda_i(t) \in [0, 1]$ for all $i \leq n_p$ and all $t \geq 0$. The number of parameters can be chosen as $2N$ (instead of the more intuitive N^2) due to the fact that $M(t)$ and $\dot{M}(t)$ are diagonal (and require only N parameters to be exactly represented).

Using (3.5) in (3.4), we get:

$$\dot{\zeta}(t) = \left(A_{\zeta_0} + \sum_{i=1}^{n_p} \lambda_i(t)A_{\zeta_i}\right)\zeta(t) + B_{lin}\widetilde{u}_p(t) \tag{3.6}$$

To reject some disturbances, the dynamical system is extended to include an output error integrator with dynamics $\dot{E} = \varepsilon \doteq -C\zeta$ (C in $\mathbb{R}^{N_c\times N}$ is the output matrix). The resulting extended matrices are, $\forall i \in \{1, 2, \ldots, n_p\}$:

$$z \doteq \begin{bmatrix}\zeta\\ E\end{bmatrix}, \quad A_0 \doteq \begin{bmatrix}A_{\zeta_0} & 0\\ -C & 0\end{bmatrix}, \quad A_i \doteq \begin{bmatrix}A_{\zeta_i} & 0\\ 0 & 0\end{bmatrix}, \quad B_e \doteq \begin{bmatrix}B_{lin}\\ 0\end{bmatrix}$$

and the dynamics of the extended system are therefore given by:

$$\dot{z}(t) = \left(A_0 + \sum_{i=1}^{n_p} \lambda_i(t)A_i\right)z(t) + B_e\widetilde{u}_p(t) \tag{3.7}$$

This model will be used in this chapter for deriving a polytopic control law. The state z belongs to $\mathbb{R}^{N_e}$, where $N_e \doteq N_c + N$.

3.2 Controller Synthesis

In order to construct the polytopic control law, we must first define the vertices of the polytope. In the case considered in this chapter the polytope is, in fact, a scaled hypercube constructed by defining the set of all partitions of $\mathscr{N}_p \doteq \{1, 2, \ldots, n_p\}$ as

$$\Omega(\mathcal{N}_p) \doteq \left\{(\mathscr{C}_j, \mathscr{D}_j) \mid \mathscr{C}_j \cap \mathscr{D}_j = \emptyset, \mathscr{C}_j \cup \mathscr{D}_j = \mathcal{N}_p\right\}.$$

The cardinality of this set is $card\ \Omega(\mathcal{N}_p) = 2^{n_p}$ (each element can be, or not, in a given subset). Each partition represents an extreme case of variation within the set of parameters (i.e. if a given element is in the partition the corresponding parameter is set at its maximum allowable value, otherwise the parameter is at its minimum allowable value). The polytopic control law will then be a convex combination of the set of vertex controllers $K_1, \ldots, K_{2^{n_p}} \in \mathbb{R}^{3\times N}$ constructed by choosing adequate coefficients $\lambda_1(t), \ldots, \lambda_{n_p}(t) \in [0, 1]$ (representing the value of each parameter within its allowable range of variation) as:

$$\tilde{u}_p(t) = \sum_{j=1}^{2^{n_p}} \beta_j(t) K_j z(t) \tag{3.8}$$

where the coefficients (positive, adding to 1) are calculated as:

$$\beta_j(t) = \prod_{k\in\mathscr{C}_j} (1 - \lambda_k(t)) \prod_{l\in\mathscr{D}_j} \lambda_l(t)$$

for the sets:

$$(\mathscr{C}_j, \mathscr{D}_j) \in \Omega(\mathcal{N}_p), \ \forall j \in \mathcal{N}_p$$

Remark 3.1 The condition that $\sum_{j=1}^{2^{n_p}} \beta_j(t) = 1$ for all $t \geq 0$ (necessary for the combination to be convex) can be shown by an induction argument on the number of parameters n_p.

The next theorem, from [3], states sufficient conditions written in terms of LMIs to compute a polytopic control law. These conditions are numerically tractable and there are many tools to find solutions (if any) that verify the inequalities and to optimize some parameters relevant for control applications. See Sect. 3.3 for an application of these numerical tools to the safety factor profile control problem.

Theorem 3.1 *A polytopic control law, as defined in* (3.8)*, that quadratically stabilizes system* (3.7) *can be constructed by setting* $K \doteq Q_j W^{-1}$*, with* $W \in \mathbb{R}^{N_e\times N_e}$ *a positive definite symmetric matrix and* $Q_j \in \mathbb{R}^{3\times N_e}$*,* $j = 1, 2, 3, \ldots, 2^{n_p}$*, full matrices such that the following LMIs are verified*[1]*:*

$$\begin{bmatrix} \varepsilon^{-1}\mathbb{I}_{N_e} & W \\ W & -M_j \end{bmatrix} \succ 0, \ \forall j \in \{1, 2, 3, \ldots, 2^{n_p}\} \tag{3.9}$$

where ε is a positive constant and, for all j, M_j is defined as:

[1] The symbol $\cdot \succ 0$ means that a matrix is positive definite.

$$M_j \doteq \left(A_0 + \sum_{i=1}^{n_p} s_{i,j} A_i\right) W + W \left(A_0 + \sum_{i=1}^{n_p} s_{i,j} A_i\right)^T + B_e Q_j + Q_j^T B_e^T$$

with, for all j, $s_{i,j} = 0$ if $i \in \mathscr{C}_j$, and $s_{i,j} = 1$ otherwise.

Sketch of Proof Consider the candidate Lyapunov function:

$$V(z) = z^T P z \tag{3.10}$$

with $P = P^T \succ 0$. Taking the derivative of V in the direction of the solution to (3.7) with control law (3.8), we obtain:

$$\dot{V} = z^T P \left(A_0 + \sum_{i=1}^{n_p} \lambda_i A_i + B_e \sum_{j=1}^{2^{n_p}} \beta_j K_j\right) z + z^T \left(A_0 + \sum_{i=1}^{n_p} \lambda_i A_i + B_e \sum_{j=1}^{2^{n_p}} \beta_j K_j\right)^T P z \tag{3.11}$$

Using the Schur complement, the inequality condition of the theorem can be rewritten as:

$$\left(A_0 + \sum_{i=1}^{n_p} s_{i,j} A_i\right) W + W \left(A_0 + \sum_{i=1}^{n_p} s_{i,j} A_i\right)^T + B_e Q_j + Q_j^T B_e^T + \varepsilon W^2 \prec 0 \quad \forall j \in \{1, 2, 3, \ldots, 2^{n_p}\} \tag{3.12}$$

Setting $W \doteq P^{-1}$ and $Q_j \doteq K_j W$, multiplying each inequality by the corresponding β_j, adding all the inequalities together, exchanging the order of summation and after some computations (involving the fact that $\sum_{j=1}^{2^{n_p}} \beta_j(t) = 1$ and $\sum_{j=1}^{2^{n_p}} \beta_j s_{i,j} = \lambda_i$), we obtain:

$$P \left(A_0 + \sum_{i=1}^{n_p} \lambda_i A_i + B_e \sum_{j=1}^{2^{n_p}} \beta_j K_j\right) + \left(A_0 + \sum_{i=1}^{n_p} \lambda_i A_i + B_e \sum_{j=1}^{2^{n_p}} \beta_j K_j\right)^T P \prec -\varepsilon \mathbb{I}_{N_e} \tag{3.13}$$

which, in turn, implies that V is a Lyapunov function for the system (3.7) under control law (3.8). For the detailed proof, see [3]. □

The next proposition is necessary to bound the gain (in the L^2 norm sense) of the polytopic controller obtained in the previous result.

Proposition 3.1 *Let $W \in \mathbb{R}^{N_e \times N_e}$ be a positive definite matrix, $K \in \mathbb{R}^{3 \times N_e}$ a full matrix and $Q \doteq KW$. A sufficient condition to guarantee that $|K|_2 < \sqrt{\gamma}$ is that the following LMIs are satisfied:*

$$\begin{bmatrix} -\mathbb{I}_3 & Q \\ Q^T & -\gamma \mathbb{I}_{N_e} \end{bmatrix} \prec 0$$
$$W \succ \mathbb{I}_{N_e}$$

where $\mathbb{I}_l$ represents the $l \times l$ identity matrix.

The proof of this proposition follows from using the Schur complement and can be found in [3].

From Theorem 3.1 and Proposition 3.1 the next corollary holds.

Corollary 3.1 *Given $\gamma > 0$, a polytopic control law as defined in* (3.8) *that quadratically stabilizes system* (3.7) *and has an L^2 gain between the state and control input strictly less than $\sqrt{\gamma}$ can be computed by setting $K_j \doteq Q_j W^{-1}$, with $W \in \mathbb{R}^{N_e \times N_e}$ a positive definite symmetric matrix and $Q_j \in \mathbb{R}^{3 \times N_e}$, $j = 1, 2, 3, \ldots, 2^{n_p}$, full matrices such that the following LMIs are verified:*

$$\begin{bmatrix} \varepsilon^{-1} \mathbb{I}_{N_e} & W \\ W & -M_j \end{bmatrix} \succ 0$$
$$\begin{bmatrix} -\mathbb{I}_3 & Q_j \\ Q_j^T & -\gamma \mathbb{I}_{N_e} \end{bmatrix} \prec 0$$
$$W \succ \mathbb{I}_{N_e}$$
$$\forall j \in \{1, 2, 3, \ldots, 2^{n_p}\}$$

where M_j is defined as in Theorem 3.1.

Remark 3.2 When trying to find a solution to this set of LMIs, the relation between the desired convergence rate ε and the gain limit γ plays a crucial role. Intuitively, it is not possible to arbitrarily accelerate the system while keeping the control gain small. Therefore, for large values of ε (with respect to a given γ), no feasible solution will exist.

3.3 Results for a Tore Supra Plasma Shot

3.3.1 Implementation

An important step required to test the approach presented in this chapter is to find a basis that adequately represents the time-varying coefficients (depending on the resistivity and its time-derivative). Although a basis of size $2N$ is enough to perfectly represent the desired coefficients (as previously mentioned), this would result in a system of 2^{2N} LMIs of size $4N_e+3$ which would be too computationally demanding. Therefore, a small basis (consisting of 5 vectors, and therefore 5 parameters) that closely approximates the desired coefficients, was chosen. The approximation error was (a posteriori) determined to be under 1 % on average for the $A_{ct}M(t)$ term and peaking around 10 % for a few points of $M^{-1}(t)\dot{M}(t)$.

In order to compute the value of the parameters at each time step, a constrained least-squares optimization problem was solved using quadratic programming. Positivity constraints were introduced to ensure that all of the vertices of the polytope have a physical meaning (i.e. to avoid negative resistivity values). Other optimization algorithms could be implemented without significant modifications (for example, a recursive least-squares algorithm). The resulting LMIs were solved using SeDuMi, see [9] and YALMIP, see [7]. These numerical tools are commonly used to solve constrained optimization problems written in terms of LMIs. See e.g. [1] for an introduction on the use of such matrix inequalities for control purposes.

3.3.2 Simulation Results

Based on the reconstructed profiles from Tore Supra shot TS-35109, a suitable reference was chosen for three points in the radial poloidal magnetic flux profile ψ. This reference was chosen similar to those in [4] and [3]. Shot TS-35109 has the following global characteristics:

- Total input power: 1.8 MA
- Total plasma current (assumed constant): 0.6 MW

The discretization follows the procedure described in [11]. Two different discretized models were used in the simulations: a low-order model ($N = 8$) used for the computation of the polytopic controller, and a higher-order model ($N = 22$) used for the simulation. The difference of order between the simulation model and the controller was chosen to better capture the effect of using a finite-dimensional controller to close the loop in an infinite-dimensional system.

The three simulations are initialized by applying the first 8 s of the open-loop control used in shot TS-35109 and then closing the loop with the polytopic controller. The operating point is changed at 20 s to test the transient response of the closed-loop system.

The behavior of the system under three different control gains is shown in Figs. 3.1, 3.2 and 3.3. The chosen LMI formulation allows for easy tuning of the controller by modifying two scalar quantities ε and γ. In these simulations, the value of gamma was fixed and then the maximum value of ε for which a solution could be found for the LMI system was chosen. The plots 3.1, 3.2 and 3.3 show the evolution of the error between the target profile and the obtained profile for ψ. It should be noted, however, that the controller only tracks three points corresponding to ψ_1, $\psi_{N/2}$ and ψ_N, (i.e.

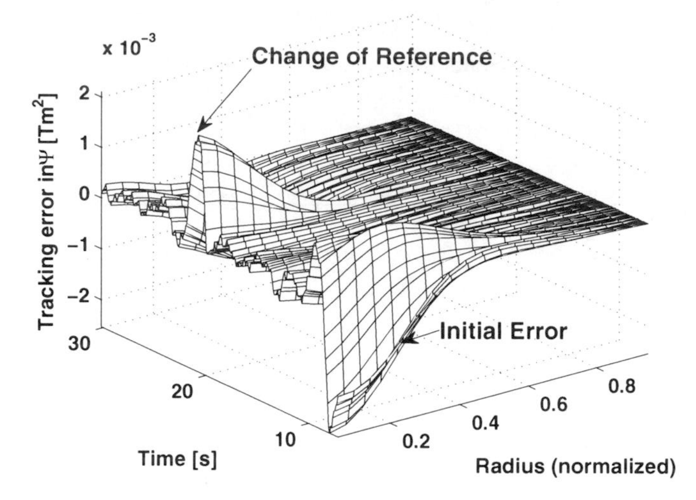

Fig. 3.1 Tracking error around $\overline{\psi}$ with LMI controller with low gain (*plain line*: numerical simulation, *dashed line*: the reference)

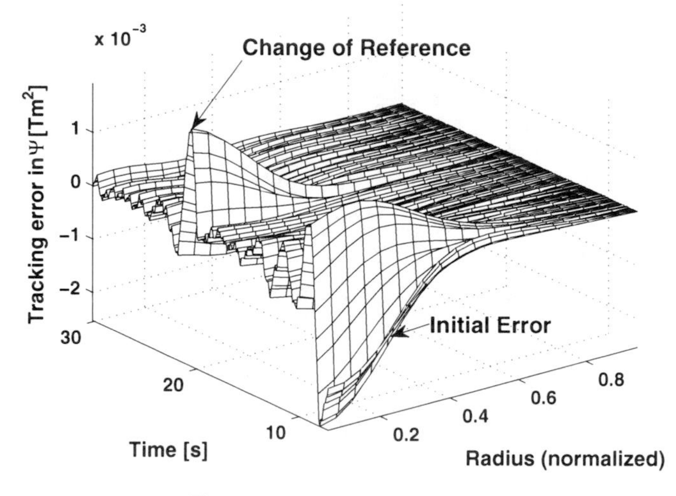

Fig. 3.2 Tracking error around $\overline{\psi}$ with LMI controller with low gain (*plain line*: numerical simulation, *dashed line*: the reference)

a point near the center of the plasma, a point at mid-radius and a point near the outer edge of the plasma). In the plots, it can be seen that, as the gain of the controller increases, the width and amplitude of the error during the initialization phase and change of reference are reduced. The plots also show that the system presents more oscillations.

It should be noted that the polytopic approach presented in this chapter requires the solution of a system of LMIs that grows exponentially with the number of parameters

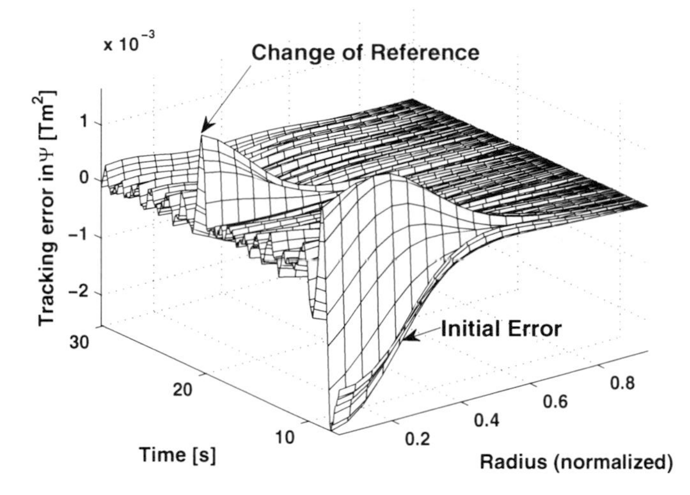

Fig. 3.3 Tracking error around $\overline{\psi}$ with LMI controller with medium gain (*plain line*: numerical simulation, *dashed line*: the reference)

chosen to represent the time-varying coefficients of the evolution equation. However, this can be done only once, offline, and takes less than a minute for the example considered.

3.4 Summary and Conclusions on the Polytopic Approach

In this section, a polytopic controller was developed for the regulation of the magnetic flux profile in a tokamak plasma based on the model presented in the previous chapter using a polytopic controller. A sufficient condition for the stability of the closed-loop system with bounded time-varying parameters and bounded time derivatives of these parameters is established. The resulting controller was then tested in simulation with a more precise model to test the robustness of the approach with respect to unmodeled dynamics, disturbances and approximation errors.

Although this approach adequately addresses some of the most important drawbacks of previous approaches like [4] (i.e. the absence of stability guarantees for the closed-loop system with time-varying matrices) it still does not entirely satisfy the requirements of the physical system. First, small and slow variations of the operating point must be considered in order to limit the norm of the $\dot{M}M^{-1}$ term. Second, the use of a linearized version of the actuator constraints limits the use of the control law far from the calculated operating point. Third, the use of three parameters in the Gaussian as control inputs is unrealistic since only two engineering parameters are available in the LH antennas (the power P_{lh} and the refractive index $N_{\|}$). Fourth, the complexity of finding a suitable control law grows exponentially with the size of the basis chosen to represent the time-varying matrices and the conservatism of the sufficient condition might make the problem unfeasible for large variations of some parameters (crucially including the time-derivative of the diffusivity coefficients). Finally, the algorithm remains computationally expensive, which may be a problem for real-time implementation.

The approach presented here thus has some important drawbacks that leave several of the key challenges presented in Chap. 2 unsolved. Furthermore, the possibility to control a few points in the poloidal magnetic flux profile does not address the main objective of controlling the safety-factor profile. Since extending these results to the gradient of the magnetic flux is not straightforward and requires imposing some limits in the variation of the resistivity profile, a different, infinite-dimensional, approach will be pursued in the rest of this book.

References

1. S. Boyd, L. El Ghaoui, E. Feron, V. Balakrishnan, Linear matrix inequalities in system and control theory. Soc. Ind. Math. **15** (1987)
2. C. Briat, Robust Control and Observation of LPV Time-Delay systems. PhD thesis, Grenoble INP, France, 2008
3. F. Bribiesca Argomedo, C. Prieur, E. Witrant, S. Brémond, Polytopic control of the magnetic flux profile in a tokamak plasma. in *Proceedings of the 18th IFAC World Congress*, Milan, Italy (2011) pp. 6686–6691
4. F. Bribiesca Argomedo, E. Witrant, C. Prieur, D. Georges, S. Brémond, Model-based control of the magnetic flux profile in a tokamak plasma, in *Proceedings of the 49th IEEE Conference on Decision and Control* (Atlanta, GA., 2010), pp. 6926–6931
5. W. Gilbert, D. Henrion, J. Bernussou, D. Boyer, Polynomial LPV synthesis applied to turbofan engines. Control Eng. Pract. **18**(9), 1077–1083 (2010)
6. D. Leith, W. Leithead, Survey of gain-scheduling analysis and design. Int. J. Control **73**(11), 1001–1025 (2000)
7. J. Löfberg, YALMIP : A toolbox for modeling and optimization in MATLAB, in *Proceedings of the CACSD Conference*, Taipei, Taiwan (2004)
8. W. Rugh, J. Shamma, Research on gain scheduling. Automatica **36**, 1401–1425 (2000)
9. SeDuMi. Sedumi website: http://sedumi.ie.lehigh.edu/, 2011
10. M.G. Wassink, M. Van de Wal, C.W. Scherer, O. Bosgra, LPV control for a wafer stage: beyond the theoretical solution. Control Eng. Pract. **13**, 231–245 (2005)
11. E. Witrant, E. Joffrin, S. Brémond, G. Giruzzi, D. Mazon, O. Barana, P. Moreau, A control-oriented model of the current control profile in tokamak plasma. Plasma Phys. Control. Fusion **49**, 1075–1105 (2007)

Chapter 4
Infinite-Dimensional Control-Lyapunov Function

In this chapter, we analyze the stability of a diffusion-like equation that evolves in a closed disk. The symmetry conditions imposed in the evolution of this equation were chosen according to the steady-state operation of the tokamak, where the evolution equation is averaged over isoflux surfaces, i.e. magnetic surfaces with the same index ρ (refer to Appendix B). This chapter recalls some previous results by the authors from [1] and [2] and develops some new results, exploring the potential use of other Lyapunov functions.

The current models for the evolution of the internal states of a tokamak present high levels of uncertainty and the online reconstruction of the variables required for a feedback control strategy is subject to errors and noise. The robustness of the closed-loop system is thus extremely important. In this chapter, we study stability in the sense of so-called *Input-to-State Stability*. This concept implies that, given a bounded input to the dynamical system, the internal states remain bounded. This notion of stability is extremely useful when dealing with physical systems, where the energy of the system remains bounded as long as the input to the system is bounded. A comprehensive survey on Input-to-State Stability (ISS) concepts for finite-dimensional systems is available in [18]. In this chapter the system considered is infinite-dimensional and some concepts may require some slight adaptation. We rely on the development of a strict Lyapunov function for the magnetic flux dynamics. For another approach to obtain ISS properties in the infinite-dimensional setting through a frequency domain approach, the reader is referred to [9].

4.1 Lyapunov Functions for Distributed Parameter Systems

As strict Lyapunov functions are at the core of this chapter, some background on Lyapunov theory for infinite-dimensional systems should be given. Being a long-standing research topic and a very active one, a comprehensive list of works dealing with Lyapunov functions (e.g. see the recent survey [4]) is outside the scope of this

F. Bribiesca Argomedo et al., *Safety Factor Profile Control in a Tokamak*,
SpringerBriefs in Control, Automation and Robotics,
DOI: 10.1007/978-3-319-01958-1_4,

book. However, a brief overview is provided by the following articles. Concerning parabolic PDEs (such as the magnetic flux dynamics in a tokamak plasma) [3] proves the existence of a global solution to the heat equation using a Lyapunov function. The series of articles [10, 16] and [17] constructs a Lyapunov function with unknown destabilizing parameters (also for the heat equation). For other classes of PDEs, [6] uses a Lyapunov function for the problem of stabilizing a rotating beam; [7] proposes a Lyapunov function to build a stabilizing boudary controller for a system of conservation laws; [5] uses Lyapunov functions to study the stability of quasilinear hyperbolic PDEs. In both [11] and [13], the use of strict Lyapunov functions to extract ISS properties is applied to both parabolic and hyperbolic PDEs. Although most of these references use simple L^2 norms as Lyapunov functions (sometimes after a change of variables), the use of weighted (or otherwise modified) L^2 norms is not new, the reader may refer to [12] and [8], just to mention some recent examples. Other related works, dealing with reaction-diffusion equations in cylindrical domains, are [19] and [21]. However, in these references, the domain does not include the central point, which precludes the use of the techniques contained in those articles.

Using a procedure similar to the one presented in the previous chapter (in the finite dimensional setting), this chapter proposes a common Lyapunov function that guarantees the stability of the system for all resistivity profiles within a given set. An important difference with respect to the previous chapter is that (with the chosen Lyapunov function) we do not constraint the rate of variation of the resistivity profiles in time. This is crucial for the safety factor regulation since the resistivity depends mostly on the temperature of the plasma, which evolves in a much faster timescale than the magnetic flux profile.

We avoid using the classical singular perturbation arguments. First, because the temperature profile depends also on the input power (which, in turn, depends on the magnetic flux profile) and on external disturbances which need not vanish or evolve on any particular timescale. Second, because the coupling between both equations is particularly complicated (i.e. not an input-output coupling through a bounded or unbounded operator but a coupling via the diffusivity coefficients, which would require a nonlinear analysis). Instead, the common Lyapunov function approach is classically used for uncertain parameter systems and particularly well suited when the parameters change rapidly (since the time-derivative of the parameters does not appear in the derivative of the Lyapunov function). Other approaches to deal with time-varying parameters can be found for example in [14, 15] and [20] but these will not be discussed here.

The construction of a strict Lyapunov function allows for an easy treatment of wide classes of disturbances, uncertainties and errors in the system. This is particularly important for a system with the degree of complexity of a tokamak. Not only the physical parameters are uncertain, but noise and biases are omnipresent in the measurements. Furthermore, the Lyapunov function constructed for the open-loop system can be treated as a control-Lyapunov function. In this case the objective is not to stabilize the system (which is open-loop stable and ISS, as will be shown in this chapter) but to accelerate the rate of convergence and attenuate the effect of disturbances. Recalling results from [1] and [2], robustness results will be given for:

- *state dynamic disturbances*;
- *actuator errors*;
- *state reconstruction and measurement errors*;
- *temperature and resistivity profile estimation errors*.

To motivate the final choice of the weighted L^2 norm as a Lyapunov function we consider candidate Lyapunov functions based on two different L^2 norms. The first one is defined for functions on the domain Ω for the Cartesian representation of (2.12)–(2.14) as:

$$\|\xi(\cdot)\|^2_{L^2(\Omega)} = \int_{\Omega} \xi^2(y)dy \tag{4.1}$$

for some function $\xi : \Omega \rightarrow \mathbb{R}$. The second one is defined for functions on an interval $[0, 1]$ based on the spatial domain defined in (2.15)–(2.17) as:

$$\|\xi(\cdot)\|^2_{L^2([0,1])} = \int_0^1 \xi^2(r)dr \tag{4.2}$$

for a function ξ now defined as $\xi : [0, 1] \rightarrow \mathbb{R}$.

It is important to note that convergence in the topology induced by the $L^2(\Omega)$ norm is not equivalent to the one obtained in the topology defined by the $L^2([0, 1])$ norm proposed in the last sections (since the $L^2(\Omega)$ norm is constructed with surface differentials which, when expressed in polar coordinates, are proportional to the radius). The choice of the $L^2([0, 1])$ norm allows us to have a non-zero weight near the center, which is necessary to regulate the central value of the safety factor.

4.2 Some Possible Lyapunov Functions

When considering the problem of guaranteeing the stability of the poloidal magnetic flux equation we may consider some Lyapunov functions that (at least in the case of constant diffusivity coefficients) simplify the analysis. We first consider the simple homogeneous case with a state $v(x, t) \doteq \psi(x, t) - \overline{\psi}(x)$. We choose to use in this section the notation v instead of the usual $\tilde{\psi}$ to emphasize the fact that we will be using the alternate boundary condition (with $V_{loop} = 0$). The dynamics of this state is given by:

$$v_t = \eta(x, t)\Delta v(x, t), \forall (x, t) \in \Omega \times [0, T] \tag{4.3}$$

where Ω is defined as in Property P_2 in Chap. 2 (an open ball of radius 1 in $\mathbb{R}^2$), with boundary condition:

$$v(x, t) = 0, \quad \forall (x, t) \in \partial\Omega \times [0, T] \tag{4.4}$$

Consider the $L^2(\Omega)$ norm of this state, defined as:

$$\|v(\cdot, t)\|^2_{L^2(\Omega)} = \int_{\Omega} v^2(y, t)dy \tag{4.5}$$

4.2.1 First Candidate Lyapunov Function

The first candidate Lyapunov function for system (4.3)–(4.4) is the following:

$$W(v, t) = \frac{1}{2} \int_{\Omega} \frac{1}{\eta(y, t)} v^2(y, t)dy \tag{4.6}$$

Remark 4.1 It should be noted that Property P_1 in Chap. 2 implies that this norm is equivalent to (4.5).

Differentiating (4.6) with respect to time along the solutions to (4.3)–(4.4), we obtain:

$$D_t W = \int_{\Omega} v(y, t)\Delta v(y, t)dy - \frac{1}{2} \int_{\Omega} \frac{\dot{\eta}(y, t)}{\eta^2(y, t)} v^2(y, t)dy$$

which, integrating by parts and using the boundary condition (4.4) implies:

$$D_t W = - \int_{\Omega} |\nabla v(y, t)|^2 dy - \frac{1}{2} \int_{\Omega} \frac{\dot{\eta}(y, t)}{\eta^2(y, t)} v^2(y, t)dy$$

Using Poincaré's inequality,

$$D_t W \leq -C_P \int_{\Omega} v^2(y, t)dy - \frac{1}{2} \int_{\Omega} \frac{\dot{\eta}(y, t)}{\eta^2(y, t)} v^2(y, t)dy$$

for some constant $C_p > 0$ depending only on the domain Ω. Using the boundedness of η we obtain:

$$D_t W \leq -2\eta_{min} C_P \left(\frac{1}{2} \int_{\Omega} \frac{1}{\eta(y, t)} v^2(y, t)dy \right) - \frac{1}{2} \int_{\Omega} \frac{\dot{\eta}(y, t)}{\eta^2(y, t)} v^2(y, t)dy$$

Defining $\alpha \doteq 2\eta_{min} C_P > 0$

$$D_t W \leq -\alpha W - \frac{1}{2} \int\limits_{\Omega} \frac{\dot{\eta}(y,t)}{\eta^2(y,t)} v^2(y,t) dy$$

and

$$D_t W \leq -\alpha W - \inf_{x \in \Omega} \left(\frac{\dot{\eta}(x,t)}{\eta(x,t)} \right) W$$

We obtain therefore that, if:

$$\inf_{(x,t) \in \Omega \times [0,T]} \left(\frac{\dot{\eta}(x,t)}{\eta(x,t)} \right) \geq -(\alpha - \varepsilon) \tag{4.7}$$

for some $\varepsilon > 0$, then W is a Lyapunov function for the system (4.3)–(4.4), with:

$$D_t W \leq -\varepsilon W \tag{4.8}$$

For the safety factor regulation, however, (4.7) is too restrictive, since it strongly limits the allowable rate of variation of the parameters at every point in time. In the next section, this Lyapunov function is modified in order to relax this condition.

4.2.2 Second Candidate Lyapunov Function

To relax condition (4.7), we use the same method as the one presented in [13] (for hyperbolic systems of conservation laws). Consider the following candidate Lyapunov function:

$$U(v,t) = e^{s_k(t)} W(v,t) \doteq \frac{1}{2} e^{\frac{1}{T} \int\limits_{t-T}^{t} \int\limits_{\tau}^{t} q_k(\xi) d\xi d\tau} \int\limits_{\Omega} \frac{1}{\eta(y,t)} v^2(y,t) dy \tag{4.9}$$

where $q_k(t) = \inf_{x \in \Omega} (\dot{\eta}(x,t)/\eta(x,t))$ and $s_k(t) \doteq \frac{1}{T} \int\limits_{t-T}^{t} \int\limits_{\tau}^{t} q_k(\xi) d\xi d\tau$.

Remark 4.2 It should be noted that, to have U equivalent to the Cartesian norm (4.5), $s_k(t)$ has to be uniformly bounded in time.

Calculating the time derivative of U along the solutions of (4.3)–(4.4), we obtain:

$$D_t U = \left(q_k(t) - \frac{1}{T} \int\limits_{t-T}^{t} q_k(m) dm \right) e^{s_k(t)} W(v,t) + e^{s_k(t)} D_t W(v,t)$$

and therefore:

$$D_t U \leq e^{s_k(t)} \left[-\frac{1}{T} \int_{t-T}^{t} \inf_{x \in \Omega} \left(\frac{\dot{\eta}(x,\tau)}{\eta(x,\tau)} \right) d\tau - \alpha \right] W(v,t)$$
$$\leq \left[-\frac{1}{T} \int_{t-T}^{t} \inf_{x \in \Omega} \left(\frac{\dot{\eta}(x,\tau)}{\eta(x,\tau)} \right) d\tau - \alpha \right] U(v,t)$$

Hence, sufficient conditions for U to be a Lyapunov function are the existence of $T, \varepsilon > 0$ and ε_1 such that:

$$\int_{t-T}^{t} \inf_{x \in \Omega} \left(\frac{\dot{\eta}(x,\tau)}{\eta(x,\tau)} \right) d\tau \geq -T(\alpha - \varepsilon), \quad \forall t \in [0,T] \tag{4.10}$$

and

$$\frac{1}{T} \int_{t-T}^{t} \int_{\tau}^{t} \inf_{x \in \Omega} \left(\frac{\dot{\eta}(x,\xi)}{\eta(x,\xi)} \right) d\xi d\tau \geq \varepsilon_1, \quad \forall t \in [0,T] \tag{4.11}$$

The first condition implies:

$$D_t U \leq -\varepsilon U, \quad \forall t \in [0,T] \tag{4.12}$$

while the second guarantees that:

$$U \geq e^{\varepsilon_1} W, \quad \forall t \in [0,T] \tag{4.13}$$

which implies the convergence of the desired norm (4.5) due to the equivalence of norms stated in Remark 4.1.

Even though this condition is less strict than (4.7) it is still too conservative for the q-profile regulation, due to the fast evolution of the diffusivity coefficients driven by the system inputs (including possible disturbances). Conditions (4.7) and (4.10) require, to guarantee the stability of the system, to either limit the rate of variation of the diffusivity coefficients (incompatible with the physical evolution of the temperature equation) or to guarantee a control gain large enough to overcome the possible positive term in the time-derivative of the Lyapunov function (in general incompatible with the constraints imposed on the actuators).

4.3 Selected Candidate Lyapunov Function and Nominal Stability

Two problems are tackled in this section:

- Constructing a Lyapunov function that captures the open-loop stability and robustness of the physical system considered (since most tokamaks to date operate in open-loop configurations, an important degree of robustness is expected).
- Selecting a Lyapunov function that can be used to build an adequate controller for the system. In this context, the open-loop system being stable, we look for more than a stabilizing controller: we require the closed loop system to retain the ISS properties of the open-loop and we use the controller to adjust the ISS gains and convergence rate.

As a first step towards developing a control Lyapunov function for the magnetic flux dynamics, we start do not constrain the shape of the admissible control input and build a Lyapunov function that captures the stability of the system in open-loop. A feedback law that assigns the exponential convergence rate of the system (2.15)–(2.17) is then proposed (assuming an unconstrained control input u). Throughout this section, we consider the normalized magnetic flux gradient z as the state and the boundary conditions are thus assumed to be homogeneous:

$$z(0,t) = z(1,t) = 0, \quad \forall t \in [0,T) \tag{4.14}$$

The error in the tracking of the plasma current is handled in two different ways, in this chapter and in the next one, in order to obtain robustness results. In the rest of this book, unless explicitly stated, we will refer to the $L^2([0,1])$ norm (defined in (4.2)) simply as the L^2 norm.

4.3.1 Selected Lyapunov Function

Consider a candidate (control) Lyapunov function given by a weighted L^2 norm as follows. Let $f : [0,1] \to (0,\infty)$ be a (strictly) positive function with bounded second derivative. Define, for system (2.15) with boundary condition (4.14) and initial condition (2.17), a weighted L^2 norm as:

$$V(z(\cdot)) = \frac{1}{2}\int_0^1 f(r) z^2(r) dr \tag{4.15}$$

It can be shown that this norm satisfies the following inequalities (and is thus equivalent to the classical L^2 norm):

$$\sqrt{\frac{f_{min}}{2}}\|z(\cdot)\|_{L^2} \leq \|z(\cdot)\|_f \leq \sqrt{\frac{f_{max}}{2}}\|z(\cdot)\|_{L^2} \tag{4.16}$$

where $f_{max} \doteq \max_{r\in[0,1]} f(r)$ and $f_{min} \doteq \min_{r\in[0,1]} f(r)$.

Remark 4.3 We use the $L^2([0, 1])$ norm to construct a candidate Lyapunov function since we can assume enough regularity in the solution of (2.15)–(2.17) for the magnetic flux dynamics. Furthermore, we avoid infinite values of the weighting function since we need this approach to be implementable. Having zero values in the weighting function is not desirable since the equivalence to the classical L^2 norm (widely used in the application) would be lost.

We can now establish a sufficient condition for the derivation of a control Lyapunov function (see also [1]):

Theorem 4.1 *If there exist a positive function $f : [0, 1] \to (0, \infty)$ with bounded second derivative and a positive constant α such that the following inequality holds:*

$$f''(r)\eta + f'(r)\left[\eta_r - \eta\frac{1}{r}\right] + f(r)\left[\eta_r\frac{1}{r} - \eta\frac{1}{r^2}\right] \leq -\alpha f(r), \quad \forall (r, t) \in [0, 1] \times [0, T) \tag{4.17}$$

then the time derivative $\dot{V}$ of the function V defined by (4.15) *verifies, for some positive constant ε_p:*

$$\dot{V} \leq -(\alpha + \varepsilon_p)V(z(\cdot, t)) + \int_0^1 f(r)\left[\eta u\right]_r z(r, t)dr, \ \forall t \in [0, T) \tag{4.18}$$

along the solutions of the PDE (2.15), *with the boundary conditions* (4.14) *and the initial condition* (2.17).

In the statement of the theorem, α could also be taken as non-negative and the sum $\alpha + \varepsilon_p$ would remain positive. This was not done since, in the rest of this book, we neglect ε_p and want α to be positive. For more information, see the remark after the proof.

Sketch of Proof Since a sufficiently regular solution to (2.15) exists, we can differentiate $V(z(\cdot, t))$ with respect to time along this solution:

$$\begin{aligned}
\dot{V} &= \int_0^1 f(r)zz_t dr \\
&= \int_0^1 f(r)\left[\eta_r u + \eta u_r\right] z dr + \int_0^1 f(r)\left(\eta_r\left[z_r + \frac{1}{r}z\right]z + \eta\left[\frac{1}{r}z_r - \frac{1}{r^2}z\right]z\right) dr \\
&\quad + \int_0^1 f(r)\eta z z_{rr} dr
\end{aligned}$$

Integrating by parts each integral term and using the homogeneous boundary conditions (4.14), the following equation is obtained:

$$\dot{V} = \int_0^1 f(r)\left[\eta_r u + \eta u_r\right] z dr - \int_0^1 f(r)\eta z_r^2 dr$$

$$+ \frac{1}{2}\int_0^1 \left(-f'(r)\eta\frac{1}{r} + f(r)\eta_r\frac{1}{r} - f(r)\eta\frac{1}{r^2} + f''(r)\eta + f'(r)\eta_r\right) z^2 dr$$

Using (4.17), we obtain the following inequality:

$$\dot{V} \leq -\alpha V(z(\cdot,t)) + \int_0^1 f(r)\left[\eta u\right]_r z dr - \int_0^1 f(r)\eta z_r^2 dr, \ \forall t \in [0,T) \qquad (4.19)$$

which, using Poincaré's inequality in the term $-\int_0^1 f(r)\eta z_r^2 dr$ to obtain the positive constant ε_p, implies the condition (4.18) of the theorem (see [1] for a more detailed proof). □

Remark 4.4 In practice, the value of ε_p that is obtained by bounding the last term in equation (4.19) using Poincaré's inequality is negligible compared to the (positive) value of α that can be obtained by solving the differential inequality using the heuristic method detailed in Appendix A. Therefore, in the rest of this book, we neglect the term ε_p.

Note that the method presented in Appendix A is not the only way to solve the differential inequality required by the theorem. In fact, in many cases a simple solution may exist. Whenever the condition:

$$\eta_r\frac{1}{r} - \eta\frac{1}{r^2} \leq -k \qquad (4.20)$$

is verified for some positive constant k, uniformly in (r,t), the regular L^2 norm verifies the conditions set in the theorem. This conditions are met, for example, when the (positive) diffusivity coefficients are monotonically non-increasing (which also includes the case when they are constant in space). The method presented in Appendix A is a practical (heuristic) way of finding suitable weighting functions for a large class of exponential-like diffusivity coefficients which are well suited for the tokamak application.

A direct consequence of Theorem 4.1 is the global exponential stability of the system considered. This result, proven in [1] is stated here without proof.

Corollary 4.1 (Global Exponential Stability) *If the conditions of Theorem 4.1 are verified, and if* $u(r,t) = 0$, *for all* (r,t) *in* $[0,1] \times [0,T)$, *then the origin of the* *system* (2.15) *with boundary conditions* (4.14) *and initial condition* (2.17) *is globally exponentially stable. The rate of convergence is $-\alpha/2$ in the topology of the norm L^2, i.e. for all $t \in [0, T)$, $\|z(\cdot, t)\|_{L^2} \leq ce^{-\frac{\alpha}{2}t}\|z_0\|_{L^2}$ for a positive constant $c \doteq \sqrt{\frac{f_{max}}{f_{min}}}$, where f_{max} and f_{min} are defined as in* (4.16).

A logical extension is to consider u as a controlled input (without adding any constraints) and build a controller that arbitrarily accelerates the closed-loop system. This is done as follows.

Corollary 4.2 (Convergence rate control) *If the conditions of Theorem 4.1 are verified, and considering $u \doteq u_{ctrl}$ where u_{ctrl} is chosen, for all $(r, t) \in (0, 1) \times [0, T)$, as:*

$$u_{ctrl}(r, t) = -\frac{\gamma}{\eta} \int_0^r z(\rho, t) d\rho \tag{4.21}$$

with $\gamma \geq 0$ a tuning parameter, then the system (2.15) *with boundary conditions* (4.14) *and initial condition* (2.17) *is globally exponentially stable. Its convergence rate is $-\beta/2 \doteq -(\alpha + \gamma)/2$, in the topology of the norm L^2.*

The proof of this corollary uses Theorem 4.1 and the fact that $[\eta\, u_{ctrl}]_r = -\gamma z$, for all $(r, t) \in [0, 1] \times [0, T)$ (see [1] for the complete proof).

4.4 Input-to-State Stability and Robustness

In the previous section, two corollaries were inferred from Theorem 4.1 . The first one adequately captured the open-loop stability of the magnetic flux gradient. The second one constructed a control law for this system that allows arbitrarily accelerating the system convergence under two strong assumptions: no uncertainties or errors exist in the system and there are no constraints in the control action available. Since this book deals with a system for which uncertainties and errors arise in almost every element in the model and the control, the robustness issue is crucial. This section makes use of the previous strict Lyapunov function to address this feedback criterion.

4.4.1 Disturbed Model

We begin by studying the effect of state disturbances and unmodeled dynamics. This analysis is crucial to deal with the issue of *pseudo-equilibria* mentioned in Chap. 2.

Consider a modified version of (2.15) that includes a state disturbance, denoted by w:

$$z_t = \left[\frac{\eta}{r}[rz]_r\right]_r + [\eta u]_r + w, \quad \forall (r, t) \in (0, 1) \times [0, T) \tag{4.22}$$

for notational simplicity, the dependence of w on (r, t) was omitted. However, the reader should be aware that this disturbance can vary in time and space.

To guarantee the existence of sufficiently regular solutions to (4.22), an extra property is assumed to hold (additionally to properties P_1–P_3 of Chap. 2):

P_4: The two-dimensional Cartesian representation of w belongs to $C^{\alpha_c,\alpha_c/2}(\overline{\Omega} \times [0, T]),\ 0 < \alpha_c < 1$.

Under these conditions, we can state the equivalent of Theorem 4.1 for (4.22) as follows:

Proposition 4.1 *Let the conditions of Theorem 4.1 hold. Then the following inequality holds:*

$$\dot{V} \leq -\alpha V(z(\cdot, t)) + \int_0^1 f(r)\,[\eta u]_r\, z dr + \int_0^1 f(r) w z dr, \quad \forall t \in [0, T) \tag{4.23}$$

along the solution of the PDE (4.22) *with the boundary conditions* (4.14) *and the initial condition* (2.17).

The proof of this theorem follows the same lines as the one of Theorem 4.1.

Proposition 4.1 allows us to formulate an ISS-like inequality for the state of the system, considering the disturbance w as an input (bounded in an L^2 sense). These inequalities are not given in the usual form for ISS since the disturbance appears inside an integral with an exponentially vanishing weight, which permits the evolution of the ISS bounds in time for time-varying disturbances. This formulation is, however, equivalent to the usual one since the time-varying disturbance can be upper-bounded by the supremum of its norm. This formulation seems to be more appropriate when the transients of the disturbance are large compared to its steady-state values.

We now formulate the main ISS result of this section.

Theorem 4.2 (ISS) *Let the conditions of Proposition 4.1 be verified and consider* $u \doteq u_{ctrl}$ *as defined in Corollary 4.2. The following inequality holds for the evolution of the PDE* (4.22) *with boundary conditions* (4.14) *and initial condition* (2.17), *for all* $t \in [0, T)$*:*

$$\|z(\cdot, t)\|_{L^2} \leq c e^{-\frac{\beta}{2}t} \|z_0\|_{L^2} + c \int_0^t e^{-\frac{\beta}{2}(t-\tau)} \|w(\cdot, \tau)\|_{L^2} d\tau \tag{4.24}$$

with $c = \sqrt{\frac{f_{max}}{f_{min}}}$, $f_{max} \doteq \max_{r\in[0,1]} f(r)$ *and* $f_{min} \doteq \min_{r\in[0,1]} f(r)$.

Sketch of Proof This proof is obtained from the following steps:

(i) Use proposition 4.1 and Corollary 4.2 to upper bound the time-derivative of the Lyapunov function along the solution of (4.22), (4.14), (2.17);

(ii) Upper bound in turn the obtained expression in terms of the product of the $\| \cdot \|_f$ norms of the disturbance and the state by applying the Cauchy-Schwarz inequality to the integral term;
(iii) Use the fact that, modulo a constant factor, the Lyapunov function is $\|z(\cdot, t)\|_f^2$ and find a bound for the time-derivative of $\|z(\cdot, t)\|_f$;
(iv) This new upper bound for the time-derivative of the norm of the state implies the desired result.

The details of this proof can be found in [1]. □

Theorem 4.2 provides a strong stability result for both the open-loop and the closed-loop systems (in the open-loop case, β reduces to α) under state disturbances. However, it is versatile enough to be applied to a wide range of errors. The next results, proven in [1], show the ease with which these can be tackled using the proposed approach.

First, we consider the input u to include both a controlled part u_{ctrl}, assumed to be governed by the previously proposed control law, and an uncontrolled part $\varepsilon_u(r, t)$ accounting for actuation errors. The result can then be stated as:

Corollary 4.3 (Actuation errors) *In addition to the conditions in Theorem 4.1, we consider* $u(r, t) \doteq u_{ctrl}(r, t) - \varepsilon^u(r, t)$, *with* u_{ctrl} *defined in Corollary 4.2 and* $\varepsilon^u(r, t)$ *a distributed actuation error verifying the regularity conditions stated in proposition* P_2. *Then, with* $w \doteq 0$, *the following inequality holds,*[1] $\forall t \in [0, T)$:

$$\|z(\cdot, t)\|_{L^2} \leq ce^{-\frac{\beta}{2}t}\|z_0\|_{L^2} + c\max\{\eta_{max}, \eta_{r,max}\}\int_0^t e^{-\frac{\beta}{2}(t-\tau)}\|\varepsilon^u(\cdot, \tau)\|_{H^1}d\tau \tag{4.25}$$

with $\eta_{max} \doteq \sup_{(r,t)\in[0,1]\times[0,T)} \mid \eta \mid$ *and* $\eta_{r,max} \doteq \sup_{(r,t)\in[0,1]\times[0,T)} \mid \eta_r \mid$.

The proof of this result follows directly replacing w by $[\eta\varepsilon^u]_r$ in Theorem 4.2.

Another important source of error is the profile reconstruction (done in real-time from the diagnostic measurements available in the tokamak). In this case, the robustness estimation can also be derived from Theorem 4.2 by writing the error term in an adequate way, which provides the following result.

Corollary 4.4 (Estimation errors in the z **profile)** *Assume that the conditions of Theorem 4.1 are verified and consider the control defined in Corollary 4.2 but substituting* z *by an estimate,* $\hat{z}(r, t) \doteq z(r, t) - \varepsilon^z(r, t)$ *for all* $(r, t) \in [0, 1] \times [0, T)$, *with* $\varepsilon^z(r, t)$ *being a distributed estimation error verifying the regularity conditions stated in proposition* P_4. *The following inequality is then verified:*

$$\|z(\cdot, t)\|_{L^2} \leq ce^{-\frac{\beta}{2}t}\|z_0\|_{L^2} + \gamma c\int_0^t e^{-\frac{\beta}{2}(t-\tau)}\|\varepsilon^z(\cdot, \tau)\|_{L^2}d\tau, \ \forall t \in [0, T) \tag{4.26}$$

[1] The H^1 norm of ξ on $[0, 1]$, is denoted as $\|\xi\|_{H^1} \doteq \|\xi\|_{L^2} + \|\frac{\partial\xi}{\partial r}\|_{L^2}$.

The error term $(\gamma\varepsilon^z)$ that appears in the system dynamics has the same form as the disturbance w in Theorem 4.2.

So far, the robustness results that have been stated guarantee (for all bounded values of disturbances and errors) the boundedness of the overall system. The next result, considering estimation errors in the resistivity profile (which arise mostly from errors in the estimation of the plasma temperature), does not guarantee the same unconditional boundedness (for the closed-loop system). This arises mostly from the fact that the η profile is used in the control law computation and multiplies the state of the system. Therefore, the norm of the equivalent control error is not uniformly bounded for an arbitrary initial condition.

Proposition 4.2 (Estimation errors in the η profile) *Assume that the conditions of Theorem 4.1 are verified and consider the control defined in Corollary 4.2 but substituting η by an estimate, $\hat{\eta}(r,t) \doteq \eta(r,t)-\varepsilon^\eta(r,t)$ for all $(r,t) \in [0,1]\times[0,T)$, with $\varepsilon^\eta(r,t)$ being a distributed estimation error verifying the regularity conditions stated in P_2. The following inequality is then verified:*

$$\|z(\cdot,t)\|_{L^2} \leq ce^{-\frac{\beta'}{2}t}\|z_0\|_{L^2}, \quad \forall t \in [0,T)$$

where $\beta' \doteq \beta + \gamma \inf_{(r,t)\in[0,1]\times[0,T)}\left(\frac{\varepsilon^\eta}{\hat{\eta}}\right) - 2\gamma c \sup_{t\in[0,T)} \|[\frac{\varepsilon^\eta}{\hat{\eta}}]_r\|_{L^2}$.

Sketch of Proof Under the conditions of the theorem, the control action is:

$$u(r,t) = -\frac{\gamma}{\hat{\eta}}\int_0^r z(\rho,t)d\rho$$

It implies that:

$$[\eta u]_r = -\gamma z - \gamma\frac{\varepsilon^\eta}{\hat{\eta}}z - \gamma\left[\frac{\varepsilon^\eta}{\hat{\eta}}\right]_r\int_0^r z(\rho,t)d\rho$$

Using the previous expression in the time-derivative of the Lyapunov function along the solution to the system, the following upper bound can be shown to hold:

$$\begin{aligned}\dot{V}(t) \leq &-\left(\beta + \gamma\left[\inf_{(r,t)\in[0,1]\times[0,T)}\left(\frac{\varepsilon^\eta}{\hat{\eta}}\right)\right]\right)V(z(\cdot,t))\\ &+ 2\gamma\|z(\cdot,t)\|_{L^2}\left\|\left[\frac{\varepsilon^\eta}{\hat{\eta}}\right]_r\right\|_f\|z(\cdot,t)\|_f\end{aligned}$$

After some computations (noticing that the product $\|z(\cdot,t)\|_{L^2}\|z(\cdot,t)\|_f$ can be written in terms of $V(z(\cdot,t))$ by the equivalence of norms):

$$\dot{V}(t) \leq -\left(\beta + \gamma \left[\inf_{(r,t)\in[0,1]\times[0,T)} \left(\frac{\varepsilon^{\eta}}{\hat{\eta}}\right)\right] - 2\gamma c \sup_{t\in[0,T)} \left\| \left[\frac{\varepsilon^{\eta}}{\hat{\eta}}\right]_r \right\|_{L^2}\right) V(z(\cdot,t)),$$
$$\leq -\beta' V(z(\cdot,t)),$$

for some positive constant c. Using the same procedure followed by the proof of Corollary 4.1 one can prove the desired result. The details of this proof can be found in [1]. □

Remark 4.5 Finding a stabilizing control law (even one that arbitrarily accelerates the convergence) for system (2.15)–(2.17) without constraint on the input is fairly simple. However, the main purpose of the results presented in Sects. 4.3 and 4.4 is the detailed account of the effect of closing the loop in terms of the rate of convergence and, much more importantly, in terms of the ISS gains of the system. This is particularly important for a system that has uncertainties and errors appearing at various places and with very different characteristics. ISS gains provide also interesting tools to estimate the effects of each separate component on the overall performance of the system. For example, from Proposition 4.2 one can conclude that if the control gain is chosen small enough, the destabilizing effect of any given level of uncertainty in the η estimation can be avoided (that is, if this level too large, that the closed-loop system will behave close to the open-loop one). This compromise is not surprising as it arises frequently when balancing performance and robustness of a system. Finally, it should be stressed that the fact that our weighted Lyapunov function captures the ISS properties of the open-loop system guarantees that stabilizing control laws can be found for the linearized system despite strong (and possibly nonlinear) shape constraints in the admissible actuation values. This property allows applying the theoretical results to the tokamak without loosing all the theoretical guarantees of stability (which would happen if the control was required to have a particular shape or amplitude, which may not be feasible given the physical constraints).

Before considering the input constraints explicitly, we will explore in the next section one way to tackle non-homogeneous boundary conditions on the system.

4.5 D^1-Input-to-State Stability

We now consider the effect of nonzero boundary conditions in the tokamak. For the tokamak, this represents errors in tracking the prescribed total plasma current I_p. This is not unrealistic since variations in the current drives' power induces transient errors in the plasma current. The first approach considered bound the norm of the state using similar techniques as those presented in previous sections.

Unlike previous sections, without introducing additional restrictions, the state will now be upper bounded not only by the magnitude of the boundary condition, but also by its time-derivative (thus the name D^1ISS, following the notation in [18]). We begin by defining some extra appropriate notations. Let $g : r \mapsto g(r)$ be an almost everywhere (a.e.) piecewise twice differentiable function, $g'(r)$ denotes an absolutely continuous function equal (a.e.) to the first derivative of g with respect to r, and g'' denotes a piecewise-continuous function equal (a.e.) to the second derivative of g with respect to r.

Remark 4.6 Henceforth, we will require the weighting function f to be almost everywhere (a.e.) twice differentiable.

4.5.1 Strict Lyapunov Function and Sufficient Conditions for D^1-Input-to-State Stability

Following the naming conventions proposed in [18] and other references, V as defined in (4.15) is said to be a strict Lyapunov function for the system (2.15)–(2.17) with homogeneous boundary conditions if, for $\varepsilon(t) \doteq -R_0\mu_0\tilde{I}_p/(2\pi) = 0$ for all $t \in [0, T)$, there exists a positive constant α such that, for every initial condition z_0 as defined in (2.17):

$$\dot{V}(t) \leq -\alpha V(z(\cdot, t)), \quad \forall t \in [0, T) \tag{4.27}$$

where, following the notation in previous chapters, $\dot{V}(t)$ represents the time derivative of V along the solution to (2.15)–(2.17).

We introduce the following assumption:

A_1: There exists a weighting function f as defined in (4.15) such that V is a strict Lyapunov function for system (2.15)–(2.17) with $u = 0$ if

$$\varepsilon(t) \doteq -R_0\mu_0\tilde{I}_P(t)/(2\pi) = 0, \quad \forall t \in [0, T)$$

which allows us to obtain the next theorem.

Theorem 4.3 **(D^1-ISS)** *Under Assumption A_1 and Properties P_1–P_3, the following inequality is satisfied, for all $t_0 \in [0, T)$, by the state of the disturbed system* (2.15)–(2.17) *with $u = 0$:*

$$\|z(\cdot, t)\|_{L^2} \leq ce^{-\frac{\alpha}{2}(t-t_0)}\left[\|z(\cdot, t_0)\|_{L^2} + \frac{1}{\sqrt{3}}|\varepsilon(t_0)|\right] + c\int_{t_0}^{t} e^{-\frac{\alpha}{2}(t-\tau)}\|\overline{\varepsilon}(\cdot, \tau)\|_{L^2}d\tau + \frac{c}{\sqrt{3}}|\varepsilon(t)| \tag{4.28}$$

where $\overline{\varepsilon}(r, t) \doteq 2\eta_r(r, t)\varepsilon(t) - r\dot{\varepsilon}(t)$, for all $(r, t) \in [0, 1] \times [t_0, T)$, and $c \doteq \sqrt{\frac{f_{max}}{f_{min}}}$ with $f_{min} \doteq \min_{r\in[0,1]}\{f(r)\}$ and $f_{max} \doteq \max_{r\in[0,1]}\{f(r)\}$.

Sketch of Proof The first step in this proof is to define a change of variables that makes the boundary condition homogeneous as follows:

$$\hat{z}(r,t) \doteq z(r,t) - r\varepsilon(t), \quad \forall(r,t) \in [0,1] \times [t_0, T) \tag{4.29}$$

The dynamics of $\hat{z}$ are then defined by:

$$\hat{z}_t = \left[\frac{\eta}{r}\left[r\hat{z}\right]_r\right]_r + 2\eta_r\varepsilon - r\dot{\varepsilon}, \quad \forall(r,t) \in (0,1) \times [t_0, T)$$

with the homogeneous boundary conditions:

$$\hat{z}(0,t) = \hat{z}(1,t) = 0, \quad \forall t \in [t_0, T) \tag{4.30}$$

and initial condition:

$$\hat{z}(r,t_0) = z(r,t_0) - r\varepsilon(t_0), \quad \forall r \in (0,1) \tag{4.31}$$

Computing the time derivative of $V(\hat{z})$ in the direction of the solution to (4.30)–(4.31) we obtain:

$$\dot{V} = \int_0^1 f(r)\hat{z}\left[\frac{\eta}{r}\left[r\hat{z}\right]_r\right]_r dr + 2\int_0^1 f(r)\hat{z}\eta_r\varepsilon dr - \int_0^1 f(r)\hat{z}r\dot{\varepsilon}dr$$

which, applying inequality (4.27) and the definition of $\overline{\varepsilon}(r,t)$ in Theorem 4.3, implies:

$$\dot{V} \leq -\alpha V(\hat{z}) + \int_0^1 f(r)\hat{z}\overline{\varepsilon}dr, \quad \forall t \in [t_0, T)$$

where $\overline{\varepsilon}(r,t)$ is uniformly bounded in $[0,1] \times [t_0, T)$. Using a procedure similar to the one used to prove Theorem 4.2 it can be shown that:

$$\|\hat{z}(\cdot,t)\|_f \leq e^{-\frac{\alpha}{2}(t-t_0)}\|\hat{z}(\cdot,t_0)\|_f + \int_{t_0}^t e^{-\frac{\alpha}{2}(t-\tau)}\|\overline{\varepsilon}(\cdot,\tau)\|_f d\tau$$

Inverting the change of variables (4.29), we obtain:

$$\begin{aligned}\|z(\cdot,t)\|_f \leq{} & e^{-\frac{\alpha}{2}(t-t_0)}\left[\|z(\cdot,t_0)\|_f + |\varepsilon(t_0)|\|r\|_f\right] \\ & + \int_{t_0}^t e^{-\frac{\alpha}{2}(t-\tau)}\|\overline{\varepsilon}(\cdot,\tau)\|_f d\tau + |\varepsilon(t)|\|r\|_f, \quad \forall t \in [t_0, T)\end{aligned}$$

which implies (4.28) and completes the proof. The full details of this proof can be found in [2]. □

Theorem 4.3 implies the following result, for the case when the boundary disturbance vanishes.

Corollary 4.5 *If there is a non-negative constant t_0 such that for all $t \geq t_0$, $\varepsilon(t)$ is zero, the state of the system* (2.15)–(2.17) *with $u = 0$ converges exponentially fast to zero in the topology of the L^2-norm.*

The next section will take advantage of the results presented in this chapter to compute a strict Lyapunov function for a set of exponential resistivity profiles. Another weight that works for a larger class of resistivity profiles is proposed in Appendix A.

4.6 Control of the Poloidal Magnetic Flux Profile in a Tokamak Plasma

4.6.1 Stability and Numerical Computation of the Lyapunov Function

To apply the theory proposed in the previous sections, an estimate of the range of variation of the plasma resistivity is required. As in [1] we consider the profile to be of the form:

$$\eta(r, t) = A(t)e^{\lambda(t)r} \tag{4.32}$$

We also consider, for simulation purposes, that the values of $A(t)$ are contained in the interval $[9.3 \times 10^{-3}, 12.1 \times 10^{-3}]$ and $\lambda(t)$ in $[4.3, 6.9]$. These limits were estimated from Tore Supra shot TS-35109 used in [22] and [1]. The weighting function f, satisfying the requirements of Theorem 4.1 was numerically calculated using Mathematica © software (Fig. 4.1).

As it was the case for the finite-dimensional polytopic approach, having a common Lyapunov function for all these profiles also guarantees the stability of the system for any convex combination of such profiles. Therefore, the actual resistivity coefficient does not necessarily need to be an exponential function. Also, since we are considering a strict Lyapunov function, $\alpha > 0$ provides a certain level of robustness with respect to both numerical errors and small deviations from the allowable set of resistivity coefficients.

The fact that we require a common Lyapunov function to hold for the whole set of diffusivity coefficients introduces some degree of conservatism. However, given that the diffusivity coefficients vary in a faster timescale than the rest of the magnetic flux dynamics, the use of a time-varying Lyapunov function may not provide a great improvement. This is due to the difficulty to find an upper-bound for the terms depending on the time-derivative of the Lyapunov function.

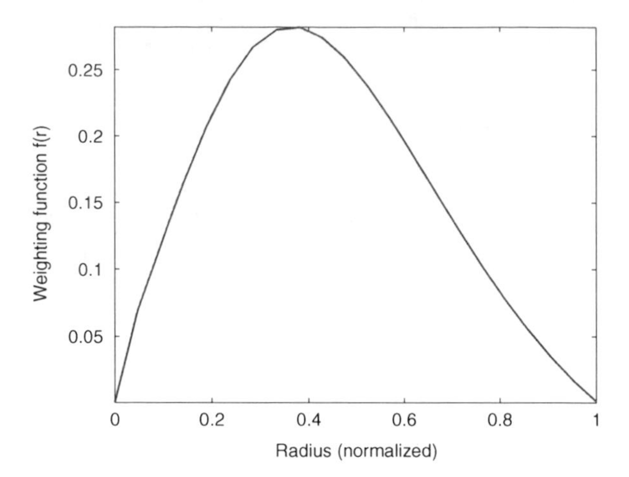

Fig. 4.1 Function f verifying the conditions of Theorem 4.1 for an exponential η with time-varying parameters. $f_{min} = 0.001$, $f_{max} = 0.2823$. ©[2013] IEEE. Reprinted, with permission, from [1]

For the particular weight used in this section, the rate of convergence obtained in simulation is faster than the one guaranteed by the theoretical results. This result could be expected, as we require the inequality in Theorem 4.1 to hold pointwise in r and the resistivity profile varies by a factor close to 1000 within the domain of r. The rate of convergence given by the numerical value of α is, however, less conservative than bounding the evolution by the lowest value of the resistivity in the plasma.

Remark 4.7 For a detailed discussion on how to compute adequate Lyapunov weights, as well as an example computed for a more general shape of diffusivity coefficients, we refer to Appendix A.

The closed-loop behavior of the system is shown in Fig. 4.3 (using the unconstrained control of Corollary 4.2) and can be compared with the response of the open-loop system presented in Fig. 4.2. The evolution of the Lyapunov function is also plotted and we can clearly see the accelerated convergence obtained with the control strategy. The controller gain was calculated using a value of $\gamma = 1.6$.

Both of these simulations were done with the same model (linear time-varying) as the one used for the control computation. In view of the safety factor regulation, we develop in the next sections more elaborated simulations involving neglected couplings and nonlinearities in the system (see [22]).

4.6.2 ISS Property and Robust Unconstrained Control of the Magnetic Flux Gradient

Building up toward the main simulations of this chapter, we first investigate the robustness by applying the unconstrained controller in Corollary 4.2 to the control-oriented model [22]. This model takes into account some nonlinearities in the evo-

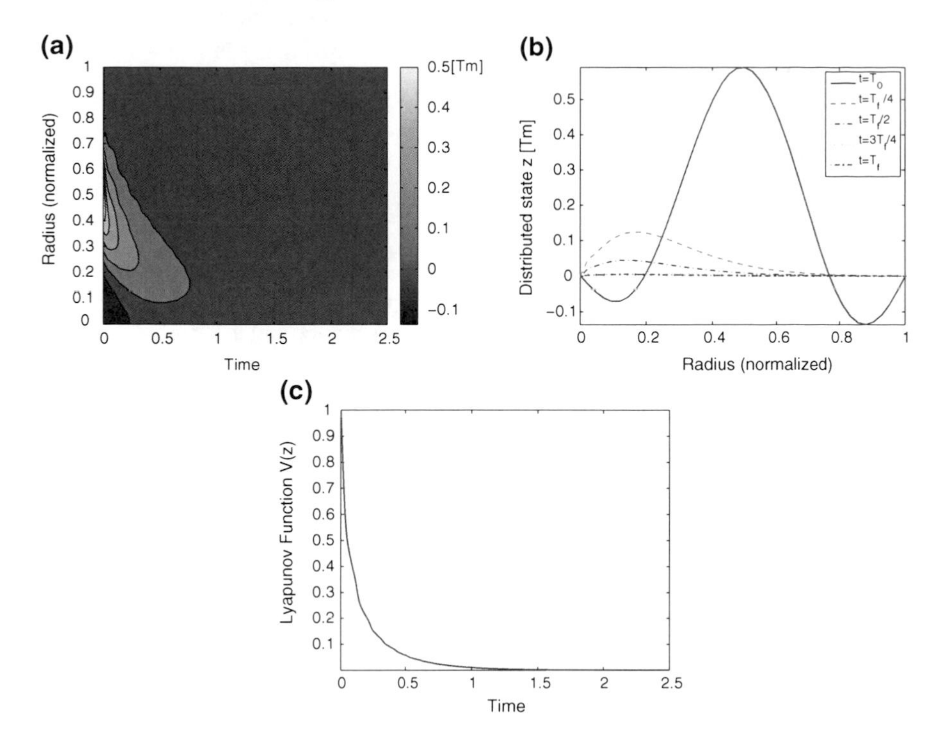

Fig. 4.2 Response of the nominal system without control action. **a** Contour plot of the solution to the PDE. **b** Time-slices of the solution to the PDE. **c** Normalized evolution of the Lyapunov function. ©[2013] IEEE. Reprinted, with permission, from [1]

lution of the magnetic flux profile, as well as the evolution of the temperature (based mostly on scaling laws). The main new effects that are introduced are:

- the bootstrap current (its variations represent a neglected nonlinearity in the evolution of the magnetic flux profile);
- other non-inductive inputs, such as ECCD (used in our test case between 8 and 20 s and acts as a disturbance);
- changes in the temperature profile (which in turns affect the resistivity profile) caused by changes in the injected power (in particular LH power variations).

These effects act as additive state disturbance (the bootstrap current variations), input disturbance (ECCD current) and multiplicative state disturbance (variations of the resistivity coefficients).

As mentioned before, the equilibrium points can be chosen from shot reconstructions. We used here a reconstructed profile data from TS-35109 in order to have an equilibrium that is physically feasible. The power injected by the ECCD antennas was chosen as being three times higher than their capacity, to emphasize their disturbance effect. Unlike for the nominal case, a lower gain was chosen with $\gamma = 0.75$

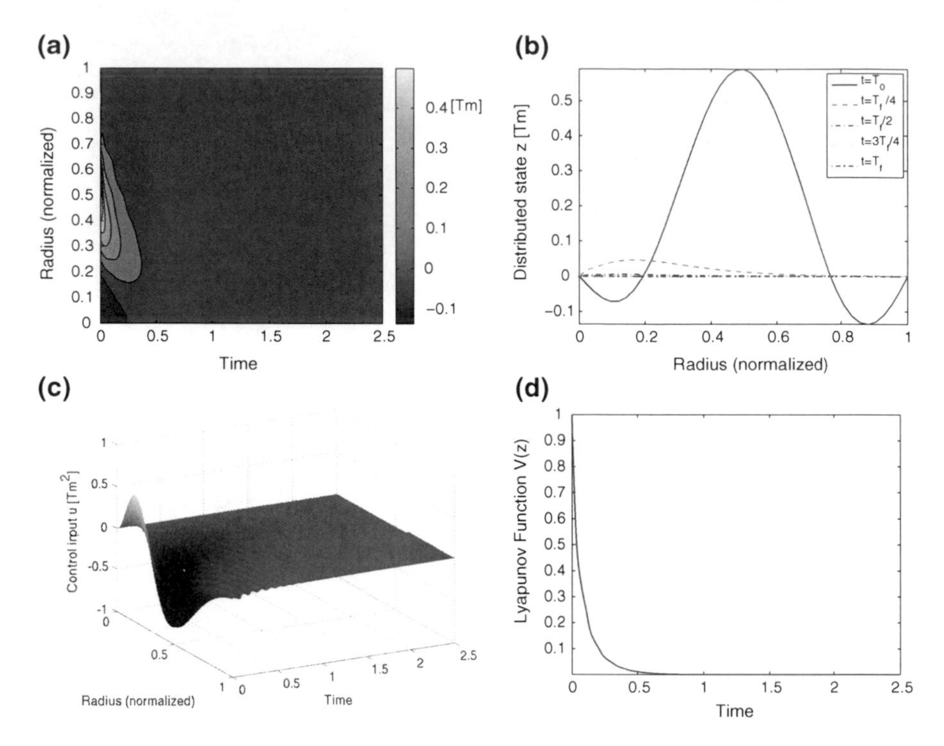

Fig. 4.3 Response of the nominal system with unconstrained control action ($\gamma = 1.6$). **a** Contour plot of the solution to the PDE. **b** Time-slices of the solution to the PDE. **c** Evolution of the control u. **d** Normalized evolution of the Lyapunov function. ©[2013] IEEE. Reprinted, with permission, from [1]

(again we encounter the performance versus robustness tradeoff mentioned in the discussion on the ISS inequalities). Two main factors limit the gain amplitude:

- the amount of noise present in the signals (tokamak signals typically have a relatively low signal-to-noise ratio, due to measurement and signal reconstruction issues, as well as the neglected fast dynamics of the particles);
- the amplitude of the control signals has to remain within some achievable physical limits.

Figure 4.4 shows the results obtained for these conditions. The control loop is closed at $t = 16$ s. Also, since these simulations were done with a more complete tokamak model, we can also show the values of the physical variables (not only the variations around an equilibrium). The magnetic flux gradient and the LH current are thus shown in Fig. 4.5. It can be appreciated that the controller both increases the rate of convergence of the system and decreases the ISS gain (the steady-state error during the ECCD disturbance is active is noticeably reduced).

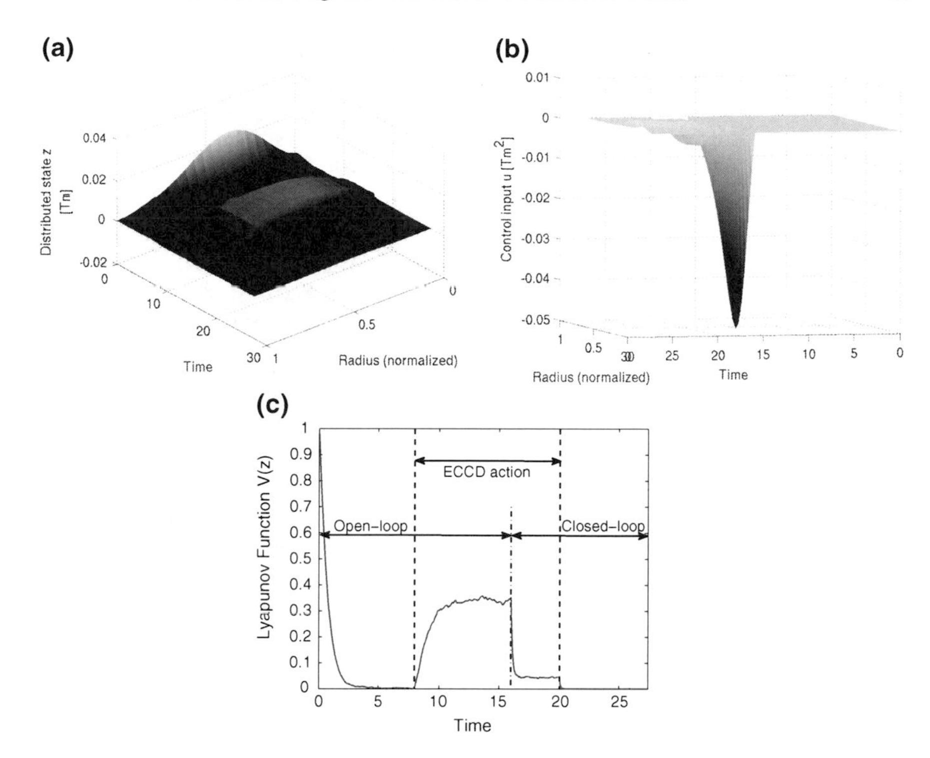

Fig. 4.4 Response of the disturbed system with a disturbance applied at $t = 8$ s and removed at $t = 20$ s. The (unconstrained) control action begins at $t = 16$ s ($\gamma = 0.75$). **a** Evolution of the z profile in time. **b** Evolution of the control input u. **c** Normalized evolution of the Lyapunov function. ©[2013] IEEE. Reprinted, with permission, from [1]

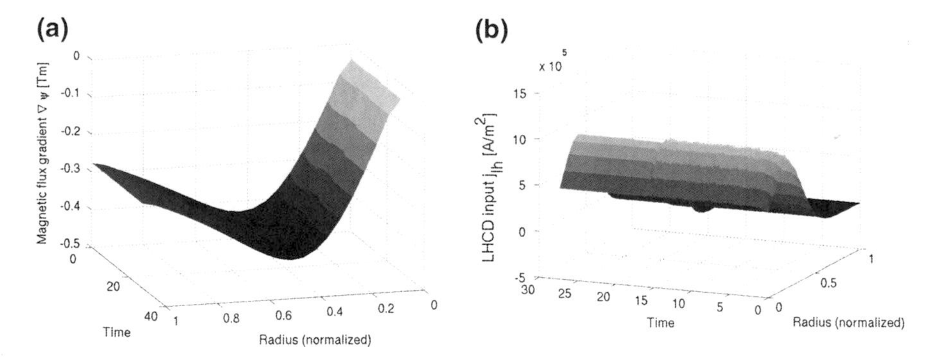

Fig. 4.5 Evolution of the physical variables when a disturbance is applied at $t = 8$ s and removed at $t = 20$ s. The (unconstrained) control action begins at $t = 16$ s ($\gamma = 0.75$). **a** Evolution of the physical $\nabla\psi$ profile. **b** j_{lh} input equivalent to the control u. ©[2013] IEEE. Reprinted, with permission, from [1]

4.6.3 Using the Lyapunov Approach to Include Actuation Constraints

The nature of the available actuators requires that we consider the effect of the shape constraints on the control action. Indeed, instead of controlling every point in the current deposit profile, only a few engineering parameters can be modified in real time. In the case of the LH antennas in Tore Supra, these parameters are the total injected power $P_{lh}(t)$ and the parallel refractive index $N_{\parallel}(t)$. For Tore Supra, two of these actuators exist and the admissible values of these parameters are:

- $0 \leq P_{lh,1} \leq 1.5\,\text{MW}$, $1.43 \leq N_{\parallel,1} \leq 2.37$
- $0 \leq P_{lh,2} \leq 3.0\,\text{MW}$, $1.67 \leq N_{\parallel,2} \leq 2.33$

Since our approach can be easily extended to other number or type of non-inductive actuators, we develop the following results as if only one set of engineering parameters $(P_{lh}, N_{\parallel})$ was available. The advantage of having a strict Lyapunov function that captures the open-loop stability of the system will become clear when looking for a feasible starting point for the optimization algorithm.

The optimization problem has two main components: the cost function and the restrictions. One restriction guarantees the closed-loop stability of the system while the other restriction constrains the gain of the controller (in order to preserve the robustness). The cost function aims to accelerate the convergence as much as possible with the available parameters (there is a link between the accelerated convergence and the ISS gain with respect to some kinds of errors). Although the cost function may not be convex (it depends mostly on the nonlinear relation between the engineering parameters and the current deposit), it can be safely assumed to be continuous. The fact that the engineering parameters take values on a compact set, together with the continuity of the cost function, guarantees both the boundedness and the existence of a minimum (which may not be unique).

We then propose to compute, at each time step, the control values $(P^*_{lh}, N^*_{\parallel})$ as the solution to the optimization problem:

$$(P^*_{lh}, N^*_{\parallel}) = \arg \min_{(P_{lh}, N_{\parallel}) \in \mathscr{U}} \int_0^1 f(r) \left[\eta u(P_{lh}, N_{\parallel})\right]_r z dr \tag{4.33}$$

subject to the constraints:

$$0 \geq \int_0^1 f(r) \left[\eta u(P^*_{lh}, N^*_{\parallel})\right]_r z dr \geq -\gamma V(z) \tag{4.34}$$

where $\mathscr{U} \doteq [P_{lh,min}, P_{lh,max}] \times [N_{\parallel,min}, N_{\parallel,max}]$ represents the (compact) set of allowable engineering parameters, and $u : \mathscr{U} \to C^{\infty}([0, 1])$ represents the variation in the LH current deposit as a function of the engineering parameters. This profile is based on scaling laws as presented in [22].

The left-hand side of the inequality (4.34) guarantees that the system verifies the conditions of Theorem 4.2 (i.e. that the system is ISS with at least the same rate of convergence as the open-loop system). The right-hand side of the inequality, in turn, was added to limit the gain of the controller. A natural question that may arise at this point is wether the optimization problem can be efficiently solved with a nonlinear and possibly non-convex cost function. The answer is that our stability proofs do not require the problem to be solved at all. As long as the engineering parameters satisfy the constraints, the closed-loop system should be stable. The cost function was added just to choose a reasonable value within the admissible set of inputs. If a different cost function and retain the same constraints, all of the above results would still hold. For example, adding a quadratic term in the variation of the engineering parameters would help avoiding long excursions of the engineering parameters around the equilibrium values.

At this point, we have proposed a method to choose a controller that satisfies the physical constraints of the actuators. We only have to show, then, that a feasible value exists at all times. It is here that the operating point comes into play. Since the evolution we are considering is around an equilibrium, choosing the parameters corresponding to the equilibrium would result in a current variation (u) equal to zero. This point satisfies both sides of the inequality (4.34). Since we are not required to find the actual solution of the optimization problem to guarantee the stability of the system, we will implement a gradient-descent algorithm (evolving on a discretized version of the parameter set) ensuring the constraints at all times. Having, at all times, a feasible starting point, the algorithm can always default to the (stable) open-loop behavior.

In order to speed the computation of the closed-loop control, the values of the function u for the chosen parameter grid were computed off-line. The algorithm only needs to do the gradient descent online. The particular version of the algorithm used for the simulations in this chapter (see also [1]) took an average of 400 μs to run when implemented as a Matlab© function on a processor running at 2.54 GHz.

4.6.3.1 Simulation Scenario: ψ_{sim}, Prescribed I_p, Constant Parameters, ECCD Disturbance

Using an equilibirum reconstructed from experimental data (Tore Supra shot TS-35109) two scenarios were constructed. The first scenario is designed to test the performance of the controller for disturbance attenuation. The chosen timestep in the simulation is 0.1 s, which is more than enough for the control algorithm to run. The disturbance is in the form of ECCD current injection (exaggerated to better illustrate the effect of the controller). The results can be seen in Fig. 4.6, where the constrained control attenuates significantly the effect of the disturbance and, once the disturbance is removed, converges close to the equilibrium (the zero of the Lyapunov function). For comparison purposes, the Lyapunov function was scaled as to have

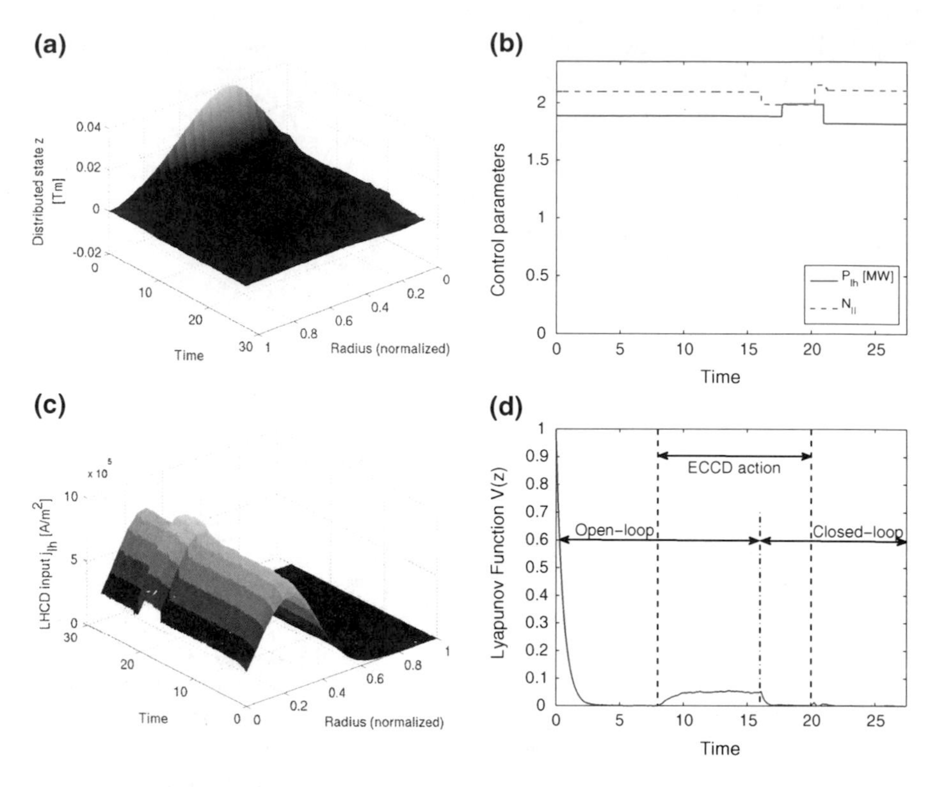

Fig. 4.6 Response of the disturbed system, disturbance applied at $t = 8$ s and removed at $t = 20$ s with constrained control action beginning at $t = 16$ s ($\gamma = 0.6$). **a** Evolution of the z profile in time. **b** Antenna parameters used to calculate the control input. **c** Evolution of the actual j_{lh} applied to the system. **d** Normalized evolution of the Lyapunov function. ©[2013] IEEE. Reprinted, with permission, from [1]

the initial value equal to 1. The second scenario was conceived to test the performance of the control when changing the operating point. The controller is activated at $t = 4$ s. The reference is changed at $t = 17$ s. The results can be seen in Fig. 4.7. The main difference between this simulation and the unconstrained one is the effect of measurement noise and the less smooth behavior of the Lyapunov function.

4.6.3.2 Simulation Scenario: ψ_{sim}, Prescribed I_p, Some P_{lh} Variation, ECCD Disturbance

Using a different equilibrium, this time reconstructed from Tore Supra shot TS-31463 involving two types of non-inductive current generation (LH and ECCD), the last scenario that will be presented in this chapter was constructed. In this case, the controller was set to track a reference that evolves in time. Only one set of equilibrium parameters is used in the controller, thus testing the robustness of the approach

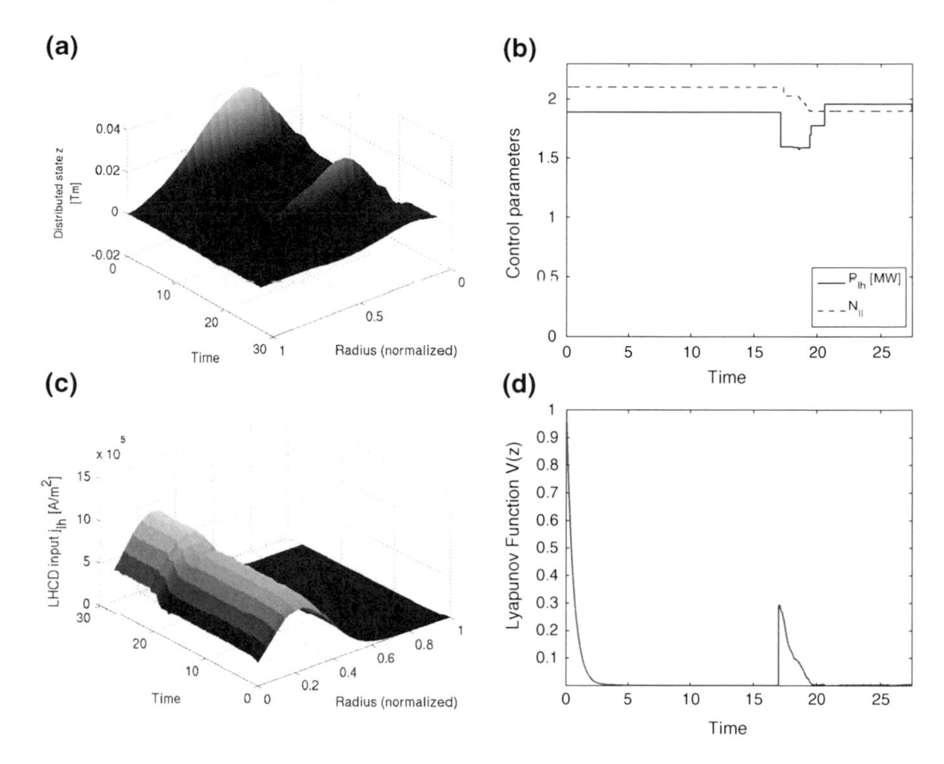

Fig. 4.7 Response of the system, change of reference applied at $t = 17$ s with constrained control action beginning at $t = 4$ s ($\gamma = 0.6$). **a** Evolution of the z profile in time. **b** Antenna parameters used to calculate the control input. **c** Evolution of the actual j_{lh} applied to the system. **d** Normalized evolution of the Lyapunov function. ©[2013] IEEE. Reprinted, with permission, from [1]

when stabilizing the system around a different operating point or when the equilibrium information is inaccurate. Figure 4.8 shows the obtained results. The controller performs well under these less-than-ideal conditions, which are natural when considering experimental implementation.

Even though these results seem promising, the degree of complexity of the tokamak cannot be adequately represented by the simplified model used for these simulations (and described in [22]). In the next chapter, it is considered coupling conditions, along with the use of a more sophisticated simulation code that incorporates different actuation models than the ones used in the control formulation.

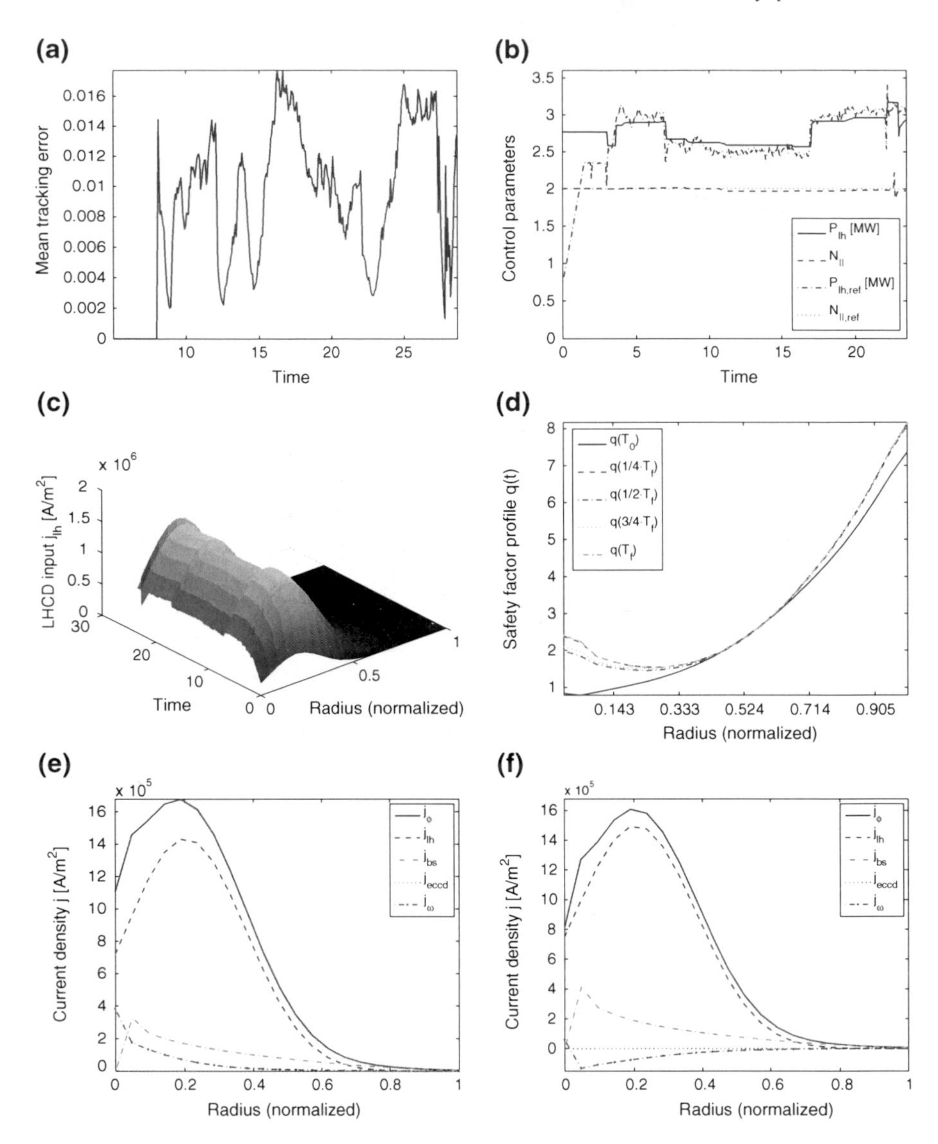

Fig. 4.8 Response of the system, with constrained control action beginning at $t = 3.1$ s ($\gamma = 2.5$). **a** Evolution of the normalized mean error in time. **b** Antenna parameters used to calculate the control input. **c** Evolution of the actual j_{lh} applied to the system. **d** Evolution of the safety factor profile. **e** Effective current profile j_ϕ and composition at $t = 14$ s. **f** Effective current profile j_ϕ and composition at $t = 19$ s. ©[2013] IEEE. Reprinted, with permission, from [1]

References

1. F. Bribiesca Argomedo, C. Prieur, E. Witrant, S. Brémond, A strict control Lyapunov function for a diffusion equation with time-varying distributed coefficients. IEEE Trans. Autom. Control **58**(2), 290–303 (2013)
2. F. Bribiesca Argomedo, E. Witrant, C. Prieur, D1-Input-to-State stability of a time-varying nonhomogeneous diffusive equation subject to boundary disturbances, in *Proceedings of the American Control Conference*, Montréal, Canada, pp. 2978–2983 (2012)
3. T. Cazenave, A. Haraux, *An Introduction to Semilinear Evolution Equations* (Oxford University Press, Oxford, 1998)
4. J.-M. Coron, *Control and Nonlinearity*, Mathematical Surveys and Monographs, vol. 136 (American Mathematical Society, Providence, 2007)
5. J.-M. Coron, G. Bastin, B. d'Andréa Novel, Dissipative boundary conditions for one-dimensional nonlinear hyperbolic systems. SIAM J. Control Optim. **47**(3), 1460–1498 (2008)
6. J.-M. Coron, B. d'Andréa Novel, Stabilization of a rotating body beam without damping. IEEE Trans. Autom. Control **43**(5), 608–618 (1998)
7. J.-M. Coron, B. d'Andréa Novel, G. Bastin, A strict Lyapunov function for boundary control of hyperbolic systems of conservation laws. IEEE Trans. Autom. Control **52**(1), 2–11 (2007)
8. A. Gahlawat, M.M. Peet, E. Witrant, Control and verification of the safety-factor profile in tokamaks using sum-of-squares polynomials, in *Proceedings of the 18th IFAC World Congress*, Milan, Italy (2011)
9. B. Jayawardhana, H. Logemann, E.P. Ryan, Infinite-dimensional feedback systems: the circle criterion and input-to-state stability. Commun. Inf. Syst. **8**(4), 413–444 (2008)
10. M. Krstic, A.T. Smyshlyaev, Adaptive boundary control for unstable parabolic PDEs. Part I. Lyapunov design. IEEE Trans. Autom. Control **53**(7), 1575–1591 (2008)
11. F. Mazenc, C. Prieur, Strict Lyapunov functions for semilinear parabolic partial differential equations. Math. Control Relat. Fields **1**, 231–250 (2011)
12. M.M. Peet, A. Papachristodoulou, S. Lall, Positive forms and stability of linear time-delay systems. SIAM J. Control Optim. **47**(6), 3237–3258 (2009)
13. C. Prieur, F. Mazenc, ISS-Lyapunov functions for time-varying hyperbolic systems of conservation laws. Math. Control Signals Syst. **24**(1), 111–134 (2012)
14. A. Smyshlyaev, M. Krstic, Closed-form boundary state feedback for a class of 1-D partial integro-differential equations. IEEE Trans. Autom. Control **49**(12), 2185–2201 (2004)
15. A. Smyshlyaev, M. Krstic, On control design for PDEs with space-dependent diffusivity or time-dependent reactivity. Automatica **41**, 1601–1608 (2005)
16. A. Smyshlyaev, M. Krstic, Adaptive boundary control for unstable parabolic PDEs. Part II. Estimation-based designs. Automatica **43**(9), 1543–1556 (2007)
17. A. Smyshlyaev, M. Krstic, Adaptive boundary control for unstable parabolic PDEs. Part III. Output feedback examples with swapping identifiers. Automatica **43**(9), 1557–1564 (2007)
18. E.D. Sontag, Input to state stability: Basic concepts and results, *Nonlinear and Optimal Control Theory* (Springer, Berlin, 2008), pp. 163–220
19. R. Vazquez, M. Krstic, Explicit integral operator feedback for local stabilization of nonlinear thermal convection loop PDEs. Syst. Control Lett. **55**, 624–632 (2006)
20. R. Vazquez, M. Krstic, Control of turbulent and magnetohydrodynamic channel flows boundary stabilization and state estimation. Syst. Control Found. Appl. (Birkhäuser, Basel, 2008)
21. R. Vazquez, M. Krstic, Boundary observer for output-feedback stabilization of thermal-fluid convection loop. IEEE Trans. Control Syst. Technol. **18**(4), 789–797 (2010)
22. E. Witrant, E. Joffrin, S. Brémond, G. Giruzzi, D. Mazon, O. Barana, P. Moreau, A control-oriented model of the current control profile in tokamak plasma. Plasma Phys. Control. Fusion **49**, 1075–1105 (2007)

Chapter 5
Controller Implementation

The main objectives of this chapter are:

- to introduce a simplified model for the evolution of the boundary condition (2.16) involving the coupling between the LH power injected into the system and the total plasma current;
- to use this simplified model to explore the possible impact of these couplings on the stability of the interconnected system;
- to implement the control law, developed in Chap. 4, in simulation using the METIS code, part of the CRONOS suite of codes [2];
- to simulate the effect of profile-reconstruction delays of 100 ms (based on the sampling time in [4]);
- to extend the control law developed in Chap. 4 in simulation using the RAPTOR code [9] for TCV scenarios.

In the first section of this chapter, we present an extended model for the system, taking into account not only the resistive diffusion equation governing the poloidal magnetic flux (infinite-dimensional system), but also the couplings that exist between the LH power injected into the system and the boundary conditions given by the total plasma current (finite-dimensional dynamical system). The dynamic behavior of the finite-dimensional subsystem is approximated using a transformer model as proposed, for instance, in [11].

Since the parameters of the LH antennas are considered as control inputs for the infinite-dimensional subsystem and are calculated considering only their impact on this subsystem, it is important to study their impact in the full interconnected system. This could be particularly challenging since in Chap. 4 we did not obtain ISS inequalities for the gradient of the magnetic flux with respect to boundary disturbances (only D^1ISS). However, with the introduction of a useful physical hypothesis (related to the total current density at the last closed magnetic surface), we are able to develop stronger (ISS) inequalities.

Two main approaches for guaranteeing the stability of the coupled system are explored in this chapter: a perfect decoupling controller for the total plasma current and a stabilizing (not decoupling) controller (both using the ohmic voltage as

F. Bribiesca Argomedo et al., *Safety Factor Profile Control in a Tokamak*,
SpringerBriefs in Control, Automation and Robotics,
DOI: 10.1007/978-3-319-01958-1_5,

actuator). The idea behind these controllers is to be able to use the constrained control law proposed in Chap. 4, with as little modifications as possible, in a more realistic (coupled) scenario while preserving some interesting theoretical properties (ISS, for instance).

For the perfect decoupling controller, a trajectory that perfectly decouples the total plasma current and the LH power and a stabilizing controller for the subsystem are required. Since constructing a stabilizing controller for this subsystem (assumed here to be LTI and verified to be controllable) is quite simple, we focus on calculating the decoupling trajectory. Eliminating the coupling, the entire system becomes nothing more than a cascade interconnection of two ISS systems. Since the output operators of both systems are bounded, this directly implies that the full system is ISS (in particular, all properties presented in Chap. 4 hold). Nevertheless, this trajectory turns out to be physically unrealistic and cannot be used for safety factor profile regulation. In particular, the presence of an integrator in the transfer between the variations of LH current ($\tilde{P}_{lh}$) and the ohmic voltage ($\tilde{V}_\Omega$) is undesirable.

Next, we turn to a stabilizing (but not perfectly decoupling) controller. Instead of calculating a perfectly decoupling trajectory, we extend the system to include an output integrator (integrating the error in the total plasma current tracking) and we assume that there exists a controller gain such that the closed-loop matrix of the system has all its eigenvalues in the left-hand side of the complex plane (with negative real parts). For an LTI system, this implies the existence of a quadratic Lyapunov function (and also the desired ISS properties of the subsystem). We then use this Lyapunov function to build a global Lyapunov candidate function (encompassing both subsystems) and find sufficient conditions for the stability of the interconnected system. Simulation results of this approach, using the METIS code, are presented and discussed in the last section.

5.1 Total Plasma Current Dynamic Model

In order to present the transformer model, we first assume the total current deposited by the LH antennas verifies the following relation $I_{lh} = \eta_{lh} P_{lh}/(R_0 \overline{n})$ (where R_0 is the location of the magnetic center, η_{lh} is a coefficient standing for the LH current drive efficiency, P_{lh} is the injected LH power and $\overline{n}$ is the electronic line average density). The current drive efficiency of the antennas can be approximated using scaling laws, such as those presented in [10].

Considering the plasma as being the secondary circuit in a transformer (with the poloidal magnetic field coils being the primary), as in [11], the evolution of $\tilde{I}_p \doteq I_p - \overline{I}_p$ around an equilibrium given by $(\overline{I}_p, \overline{P}_{lh}, \overline{N}_\parallel, \overline{V}_\Omega, \overline{I}_\Omega)$ is given by:

$$\begin{bmatrix} L_p & M \\ M & L_\Omega \end{bmatrix} \begin{bmatrix} \dot{\tilde{I}}_p \\ \dot{\tilde{I}}_\Omega \end{bmatrix} = \begin{bmatrix} -R_p & 0 \\ 0 & -R_\Omega \end{bmatrix} \begin{bmatrix} \tilde{I}_p \\ \tilde{I}_\Omega \end{bmatrix} + \begin{bmatrix} \frac{\eta_{lh} R_p}{\overline{n} R_0} & 0 \\ 0 & 1 \end{bmatrix} \begin{bmatrix} \tilde{P}_{lh} \\ \tilde{V}_\Omega \end{bmatrix} \tag{5.1}$$

where, following the notation in the rest of this book, a tilde represents the difference between the actual value and the equilibrium value of a variable (i.e. $\tilde{\xi} \doteq \xi - \overline{\xi}$). L_p and L_Ω stand for the plasma and coil inductances, respectively, M represents the mutual inductance, and R_p and R_Ω stand for the plasma and coil electrical resistances. I_Ω and V_Ω represent the ohmic current and voltage, respectively.

We further assume the initial condition is:

$$\tilde{I}_p(0) = \tilde{I}_\Omega(0) = 0 \tag{5.2}$$

5.1.1 *Perfect Decoupling and Cascade Interconnection of ISS Systems*

The first idea that we explore is the construction of a decoupling feedback (in this section referred to as a decoupling trajectory since, due to the linearity of the system, it defines a trajectory for the state variables in the finite-dimensional subsystem that can be superposed to that of an independent controller for the same subystem). The purpose behind this decoupling trajectory is to simplify the stability analysis of the system and to be able to use the control algorithm already developed in the previous chapter with few, if any, modifications.

If this perfect decoupling can be achieved, the stability analysis of the overall system is reduced to show that the cascade interconnection of two ISS systems is asymptotically stable. This result is straightforward, since the ISS property is equivalent to a bounded gain between the input and state of the system. The first system being linear and finite-dimensional, ISS implies BIBO stability (the output operator is clearly bounded by the norm of the state). The bounded output of the first system becomes a bounded input to the infinite dimensional system (via the boundary condition) and, introducing a further physical hypothesis (on the total current density at the last closed magnetic surface), the D^1ISS inequality presented on the previous chapter for boundary disturbances can be replaced by a stronger ISS one.

Consider then a *decoupling* trajectory $(\overline{\overline{V}}_\Omega, \overline{\overline{I}}_\Omega)$ calculated as:

$$\begin{bmatrix} \dot{\overline{\overline{I}}}_\Omega \\ \dot{U} \end{bmatrix} = \begin{bmatrix} -\frac{L_p R_\Omega}{L_p L_\Omega - M^2} & \frac{L_p}{L_p L_\Omega - M^2} \\ 0 & 0 \end{bmatrix} \begin{bmatrix} \overline{\overline{I}}_\Omega \\ U \end{bmatrix} + \begin{bmatrix} \frac{\eta_{lh} R_p}{\overline{n} M R_0} \\ \frac{R_\Omega \eta_{lh} R_p}{M \overline{n} R_0} \end{bmatrix} \tilde{P}_{lh}$$

$$\begin{bmatrix} \overline{\overline{V}}_\Omega \\ \overline{\overline{I}}_\Omega \end{bmatrix} = \begin{bmatrix} 0 & 1 \\ 1 & 0 \end{bmatrix} \begin{bmatrix} \overline{\overline{I}}_\Omega \\ U \end{bmatrix} + \begin{bmatrix} \frac{\eta_{lh} L_\Omega R_p}{\overline{n} M R_0} \\ 0 \end{bmatrix} \tilde{P}_{lh} \tag{5.3}$$

with initial conditions $\overline{\overline{I}}_\Omega(0) = U(0) = 0$. Here, U is merely an internal variable used to generate the required trajectory (i.e. it does not necessarily have a physical meaning). Adding this trajectory to the original equilibrium (using Eq. (5.1)) effectively decouples the variables $\tilde{P}_{lh}$ and $\tilde{I}_p$, i.e. after some computations we obtain:

$$\begin{bmatrix} L_p & M \\ M & L_\Omega \end{bmatrix} \begin{bmatrix} \dot{\tilde{I}}_p \\ \dot{\tilde{I}}_\Omega - \dot{\overline{\overline{I}}}_\Omega \end{bmatrix} = \begin{bmatrix} -R_p & 0 \\ 0 & -R_\Omega \end{bmatrix} \begin{bmatrix} \tilde{I}_p \\ \tilde{I}_\Omega - \overline{\overline{I}}_\Omega \end{bmatrix} + \begin{bmatrix} 0 \\ 1 \end{bmatrix} \left(\tilde{V}_\Omega - \overline{\overline{V}}_\Omega \right) \quad (5.4)$$

and therefore, the evolution of $\tilde{I}_p$ does not depend on $\tilde{P}_{lh}$.

Since (5.3) describes an LTI system, the transfer function between $\tilde{P}_{lh}$ and $\overline{\overline{V}}_\Omega$ can be easily calculated and it is actually a PI gain. Therefore, in order to perfectly decouple (at all frequencies) the total plasma current from the variations of the LH power, the ohmic voltage has to integrate the deviation from the equilibrium total LH current deposit. As mentioned in the chapter introduction, this decoupling feedback poses two main problems: first, small variations on the LH power will eventually result in actuator saturation (V_Ω is limited, as is the total flux that can be produced using the inductive actuators); and second, it requires a very aggressive control action that, as discussed in the next subsection, is not really required by the application. Furthermore, the total plasma current controllers currently employed are not designed to achieve perfect decoupling. Based on these observations, although this approach may be appealing from a theoretical standpoint, it is not pursued in the next sections. Instead, a second approach (with less stringent conditions on the total plasma current tracking) will be presented and we will show that, under certain conditions, we do not require such a decoupling to ensure the stability of the interconnected system.

5.1.2 Interconnection Without Perfect Decoupling

In order to deal with the shortcomings of the previous approach (thus obtaining something more easily implementable) we consider an imperfect decoupling and establish some sufficient conditions for the interconnected system to remain stable (and, under certain circumstances, with a similar rate of convergence as that obtained in the previous section). This will be assumed to be the case for the remainder of this book.

Let us define the following matrices:

$$A \doteq \begin{bmatrix} -\frac{L_\Omega R_p}{L_p L_\Omega - M^2} & \frac{M R_\Omega}{L_p L_\Omega - M^2} & 0 \\ \frac{M R_p}{L_p L_\Omega - M^2} & -\frac{L_p R_\Omega}{L_p L_\Omega - M^2} & 0 \\ -1 & 0 & 0 \end{bmatrix},$$

$$B \doteq \begin{bmatrix} -\frac{M}{L_p L_\Omega - M^2} \\ \frac{L_p}{L_p L_\Omega - M^2} \\ 0 \end{bmatrix},$$

$$D \doteq \begin{bmatrix} \frac{L_\Omega \eta_{lh} R_p}{(L_p L_\Omega - M^2)\overline{n} R_0} \\ -\frac{M \eta_{lh} R_p}{(L_p L_\Omega - M^2)\overline{n} R_0} \\ 0 \end{bmatrix}$$

Writing (5.1) as a system of ODEs, after substracting the equilibrium and extending the system to include an integrator (similar to what was done in Chap. 3) we obtain:

$$\dot{\zeta} = A\zeta + B\tilde{V}_{\Omega} + D\tilde{P}_{lh} \tag{5.5}$$

where the state of this system is defined as:

$$\zeta \doteq \left[\tilde{I}_p \; \tilde{I}_{\Omega} \; E\right]^T$$

and the extended state E represents the integral of $\tilde{I}_p$.

In this chapter, we consider only the case of constant matrices A, B, and D. Other methods, like the polytopic LPV approach shown in Chap. 3 (from [5]), could handle the time-varying system, however this would only complicate the calculations in this chapter and decrease its readability.

Given that $L_pL_{\Omega} - M^2 > 0$, the matrix A has two stable eigenvalues (with negative real part) and one marginally stable eigenvalue (corresponding to the integrator).

The nonhomogeneous boundary condition in (2.16) can be written in terms of an output of this finite-dimensional system as:

$$z(1,t) = C\zeta \tag{5.6}$$

with the output matrix defined as $C \doteq \left[-\frac{R_0\mu_0}{2\pi} \; 0 \; 0\right]$.

5.2 Modified Lyapunov Function

Following [7], we consider the following candidate Lyapunov function:

$$\begin{aligned} V(z,\zeta) &= V_1(z) + \kappa V_2(\zeta) \\ V_1(z) &= \frac{1}{2}\int_0^1 f(r)z^2(r,t)dr \\ V_2(\zeta) &= \frac{1}{2}\zeta^T P\zeta \end{aligned} \tag{5.7}$$

with $f : [0,1] \to [\varepsilon, \infty)$ a twice continuously differentiable positive function, $P = P^T \in \mathbb{R}^{2\times 2}$ a symmetric positive definite matrix, and κ a positive constant.

We now present the main result of this chapter. Although the theorem seems similar to that in [7], there are several important differences between both results (item 2. is different and different properties are used in the proof). We will therefore provide the complete proof instead of the customary *Sketch of Proof*.

Theorem 5.1 *If the following two conditions are satisfied:*

1. *there exists a twice continuously differentiable positive weight* $f : [0, 1] \to [\varepsilon, \infty)$ *as defined in* (4.15) *such that the function* V_1 *in* (5.7) *is a strict Lyapunov function for system* (2.15)–(2.17) *with homogeneous boundary conditions (i.e. verifying for some positive constant* α_1, $\dot{V}_1 \leq -\alpha_1 V_1$*);*
2. *an independent control loop (i.e. a local controller that does not take into account the infinite-dimensional dynamics when computing the control action) regulates the total plasma current while ensuring, for some symmetric positive definite matrix* $P \in \mathbb{R}^{3\times 3}$*, some matrix* $K \in \mathbb{R}^{1\times 3}$ *and some positive constant* α_2*:*

$$P[A + BK + \alpha_2 \mathbb{I}_3] \prec 0 \tag{5.8}$$

where $\cdot \prec 0$ *denotes the negative definiteness of a square matrix,* $\mathbb{I}_3$ *is the* 3×3 *identity matrix;*

then there exists κ *large enough such that the function* $V(z, \zeta)$ *is a strict Lyapunov function for the interconnected system whose derivative along the solutions of* (2.15)–(2.17) *and* (5.5) *satisfies, for some positive constants* c_1, c_2*:*

$$\dot{V}(t) \leq -c_1 V(z, \zeta) + c_2 \tilde{P}_{lh}^2(t), \forall t \in [0, T) \tag{5.9}$$

Proof Given an adequate weighting function $f(r)$ satisfying condition (i) of Theorem 3.9 we have along the solution of (2.15)–(2.17):

$$\dot{V}_1(t) \leq -\alpha_1 V_1(z) - \frac{1}{2}\left(f(1) + f'(1)\right)\eta(1, t) z^2(1, t) \tag{5.10}$$

for some $\alpha_1 > 0$. This equation is obtained by keeping the terms that depend on the boundary condition in the proof of Theorem 4.1 and assuming that the total current density, defined as in [3]:

$$j_T = -\frac{1}{\mu_0 R_0 a^2 r}\left(r\psi_{rr} + \psi_r\right)$$

is zero on the last closed magnetic surface. If avoiding this hypothesis were desirable, uniform boundedness and uniform Lipschitz-continuity in time could be assumed building upon [6], with more conservative bounds (D^1ISS instead of ISS).

Differentiating V_2 along the solution of (5.5), we get:

$$\dot{V}_2 = \frac{1}{2}\zeta^T P\left[A\zeta + B\tilde{V}_\Omega + D\tilde{P}_{lh}\right] + \frac{1}{2}\left[\zeta^T A^T + \tilde{V}_\Omega^T B^T + \tilde{P}_{lh}^T D^T\right] P\zeta \tag{5.11}$$

Defining $\tilde{V}_\Omega \doteq K\zeta$, with K a row vector of size 3, the previous equation implies:

$$\dot{V}_2 = \zeta^T P\,[A + BK]\,\zeta + \zeta^T P\left[D\tilde{P}_{lh}\right] \tag{5.12}$$

Applying Young's inequality, for any positive constant ε:

$$\dot{V}_2 \leq \zeta^T P \left[A + BK\right] \zeta + \frac{\varepsilon}{2} \zeta^T P D D^T P \zeta + \frac{1}{2\varepsilon} \tilde{P}_{lh}^2 \tag{5.13}$$

Let us choose ε small enough so that:

$$-\alpha_2 P + \frac{\varepsilon}{2} P D D^T P \prec -\alpha_3 P \tag{5.14}$$

with $0 < \alpha_3 < \alpha_2$. Then:

$$\dot{V}_2 \leq -\alpha_3 \zeta^T P \zeta + \frac{1}{2\varepsilon} \tilde{P}_{lh}^2 \tag{5.15}$$

Using the boundary condition (5.6), and the definition of C, the derivative of V along the solution of the coupled system is bounded by:

$$\begin{aligned} \dot{V}(t) \leq & -\alpha_1 V_1(z) - \frac{R_0^2 \mu_0^2}{8\pi^2} \left(f(1) + f'(1)\right) \eta(1,t) \zeta^T \zeta \\ & - 2\kappa \alpha_3 V_2(\zeta) + \frac{\kappa}{2\varepsilon} \tilde{P}_{lh}^2(t) \end{aligned} \tag{5.16}$$

We now consider two cases: depending on the sign of $f(1) + f'(1)$. We will focus first on the case where $f(1) + f'(1) < 0$. In this case, we can choose κ large enough so that:

$$-\kappa \alpha_3 P - \frac{R_0^2 \mu_0^2}{8\pi^2} \left(f(1) + f'(1)\right) \eta(1,t) \mathbb{I}_3 \prec -\frac{\alpha_4}{2} P$$

for some $0 \prec \alpha_4 \prec \alpha_3$ uniform in time (by using P_1).

The case where $f(1) + f'(1) >= 0$ is even simpler, since we can choose $\kappa = 1$ and get the same inequality with $\alpha_4 \geq \alpha_3$.

This implies that, in both cases:

$$\dot{V} \leq -\alpha_1 V_1 - \alpha_4 V_2 + \frac{\kappa}{2\varepsilon} \tilde{P}_{lh}^2 \tag{5.17}$$

and therefore:

$$\dot{V} \leq -\min\left\{\alpha_1, \alpha_4\right\} V + \frac{\kappa}{2\varepsilon} \tilde{P}_{lh}^2 \tag{5.18}$$

thus completing the proof. □

Remark 5.1 Based on Eq. (5.18) and the boundedness of $\tilde{P}_{lh}$ ($\tilde{P}_{lh}$ is bounded since both P_{lh} and $\overline{P}_{lh}$ belong to the same bounded interval in $\mathbb{R}_{\neq}^{+}$), for any bounded initial state, the state of the system will remain bounded at all times. Furthermore, the ISS gain of the system can be bounded by a function of the rates of convergence of both subsystems and the constants κ and ε used in the proof. In order to obtain as small an ISS gain as possible, it would be desirable to have a big value for ε and a small value for κ, together with as big as possible convergence rate (given by $\min\{\alpha_1, \alpha_4\}$). To obtain this, an α_2 as large as possible is desirable (which is quite logical, the faster the finite dimensional system converges, the less impact the coupling will have). The limiting case would, of course, be the case of perfect decoupling mentioned in the previous section.

Remark 5.2 From Eq. (5.18), we conclude that if, in addition to the constraints already included in the control law, see (4.34), an additional constraint is imposed as:

$$\tilde{P}_{lh}^2 \leq (-\alpha_5 + \min\{\alpha_1, \alpha_4\}) \frac{2\varepsilon}{\kappa} V(z, \zeta), \ \forall t \in [0, T] \tag{5.19}$$

for some $0 < \alpha_5 < \min\{\alpha_1, \alpha_4\}$, then the interconnected system is exponentially stable.

This condition is indeed a small-gain condition involving the ISS gain of the cascade interconnection of both subsystems, represented by $\kappa/(2\varepsilon \min\{\alpha_1, \alpha_4\})$ (see Eq. (5.18)), and the gain of the controller (given by the previous inequality).

Remark 5.3 We may also remark that if $\tilde{I}_p$ ever converges to zero, the infinite-dimensional state will converge to zero as well (this will happen, for instance when $\tilde{P}_{lh}$ is constant, due to the added integrator in the system).

5.3 Simulation Result: Closed-Loop Tracking Using METIS

In this section, instead of using the simplified control-oriented model presented in [13] for simulation purposes, we will validate the proposed approach using a more comprehensive simulation code. A simulation platform was developed using a Simulink© interface developed by the CEA (France) for the METIS code, see [1]. An integral part of this platform was the versatility to simulate the evolution of the magnetic (and other) variables in a Tokamak using different boundary conditions and actuator models (in order to test, with increasing levels of detail, the control approach presented in this book). For more details on this platform, please refer to [7].

5.3.1 General Description

In this section, two simulation cases are presented. As in previous chapters, the controller requires an approximate equilibrium in order to function. The equilibrium given to the controller will be obtained from profiles reconstructed from experimental data of Tore Supra shot TS-31463. During the test cases, however, the operating point will not correspond to the equilibrium given to the controller. In fact, both simulation cases were chosen in order to take the system far from the controller equilibrium to test its robustness and its applicability under more realistic conditions.

Although both engineering parameters can be modified by the algorithm, we built the references by varying first the LH power P_{lh} (in the first case) and the refractive index $N_{\|}$ (in the second case). As mentioned before, the range of variation of these two parameters will be chosen large enough to properly illustrate a wide set of operating conditions. In all cases, the control law is given by (4.33) and the weighting function was calculated following the procedure outlined in Appendix A.

Compared to the results presented in Chap. 4, in this case, more attention was given to the parameter tuning in order to limit the impact of measurement noise (present in the experimental data used to feed the simulation). Also, the rate of variation of the engineering parameters was limited in order to better represent the physical system.

The following information was provided to the controller:

- An estimation of the equilibrium profiles along with a reference (current and magnetic flux gradient): in this case, this estimation was deliberately chosen from at a different operating point;
- a real-time (or near-real-time) estimation of the gradient of the magnetic flux (the effect of the profile reconstruction delay will be studied later in this chapter, under simulation);
- a real-time estimation of the $\eta_{\|}$ profile;
- an estimated shape for the current deposit profile based on scaling laws, as described in [13] (in this case, the shape function was evaluated in a grid in the parameter space in order to speed the online optimization).

Since adequate scaling laws exist describing the current deposit profiles for most tokamaks and an estimated equilibrium profile can be obtained either from simulation or experimental data, these two do not represent a major problem. The magnetic flux profile reconstruction is more difficult to obtain but it is available in Tore Supra in near real time thanks to the Equinox code [4]. However, the controller proves to be robust with respect to estimation errors in both the resistivity profile and the magnetic flux profile. The estimation of plasma parameters can also be done using the method described in [9], which was used in the next section.

The only outputs from the controller are:

- The Lower Hybrid power P_{lh};
- The parallel refractive index $N_{\|}$.

An independent controller is active throughout the shot regulating the total plasma current I_p using the poloidal field coils. Every shot starts in an open-loop configuration and then, at $t = 9$ s the proposed controller is activated.

The general parameters of the flat-top phase of the shot are summarized in Table 5.1.

These global parameters were chosen similar to the ones used in [7] for easy comparison. The general structure of the shots is also similar, although the ranges and shapes of the input parameters were chosen differently. Also, the shape and magnitude of the ICRH disturbances are different (for example, in the first simulation

Table 5.1 Global parameters used for simulation, based on Tore Supra shot TS-31466, with extended range

Global shot parameters	
$\bar{n}$	1.45×10^{19} m $^{-2}$
$P_{lh,reference}$	between 1 and 3 MW
$N_{\parallel,reference}$	between 1.8 and 2.2
$B_{\phi 0}$	3.69 T
I_p	580 KA

scenario, the ICRH power is not constant during the disturbance window). In this case, the lower values of the ICRH magnitudes allow for a better rejection of the disturbances. Also, some tuning was done on the gradient descent algorithm which explains the somewhat different response of the system.

5.3.2 Simulation Scenario: METIS, Independent I_p Control, Large Variations of P_{lh}, ICRH Heating Disturbance

The test case shown in this section was constructed to replicate, as closely as possible, under simulation, the main difficulties that would be encountered in an experimental setting. First, the total plasma current cannot be perfectly controlled. Although the poloidal field coils can effectively generate current to offset the current changes generated by the variations in the non-inductive sources, the tracking cannot be perfect. Second, the equilibrium profiles given to the controller cannot be more than approximations since they have to be either obtained from simulations or reconstructed from experimental data (neither of which is completely reliable). Third, the reference given to the controller may not be achievable through the constrained actuators available. Fourth, the current deposit profiles given to the controller and those applied to the Tokamak are not the same, they are just approximations using scaling laws. Fifth, measurement noise is present in all measurements in the tokamak. All of these sources of error are integrated into the simulations in different ways:

- The tracking error in the total plasma current appears as a boundary disturbance in (2.16).
- A single equilibrium profile is given to the controller, corresponding to a point with $P_{lh} = 2.76$ MW and $N_{\parallel} = 2$.
- A (mostly temperature) disturbance is introduced with $P_{ICRH} = 1.5$ MW (ICRH).
- The LH actuator model used for simulation uses scaling laws based on parameters of the plasma and the LH waves (see [10]) to compute the efficiency and then Landau absorption, accessibility and caustics to determine the current deposit profile (see [1]).

Figure 5.1 shows the tracking performance and the control outputs during the simulation. The initial, open-loop ramp-up phase is not shown. We focus, instead, on the flat-top evolution of the safety factor profile. For clarity, Fig. 5.1a shows only

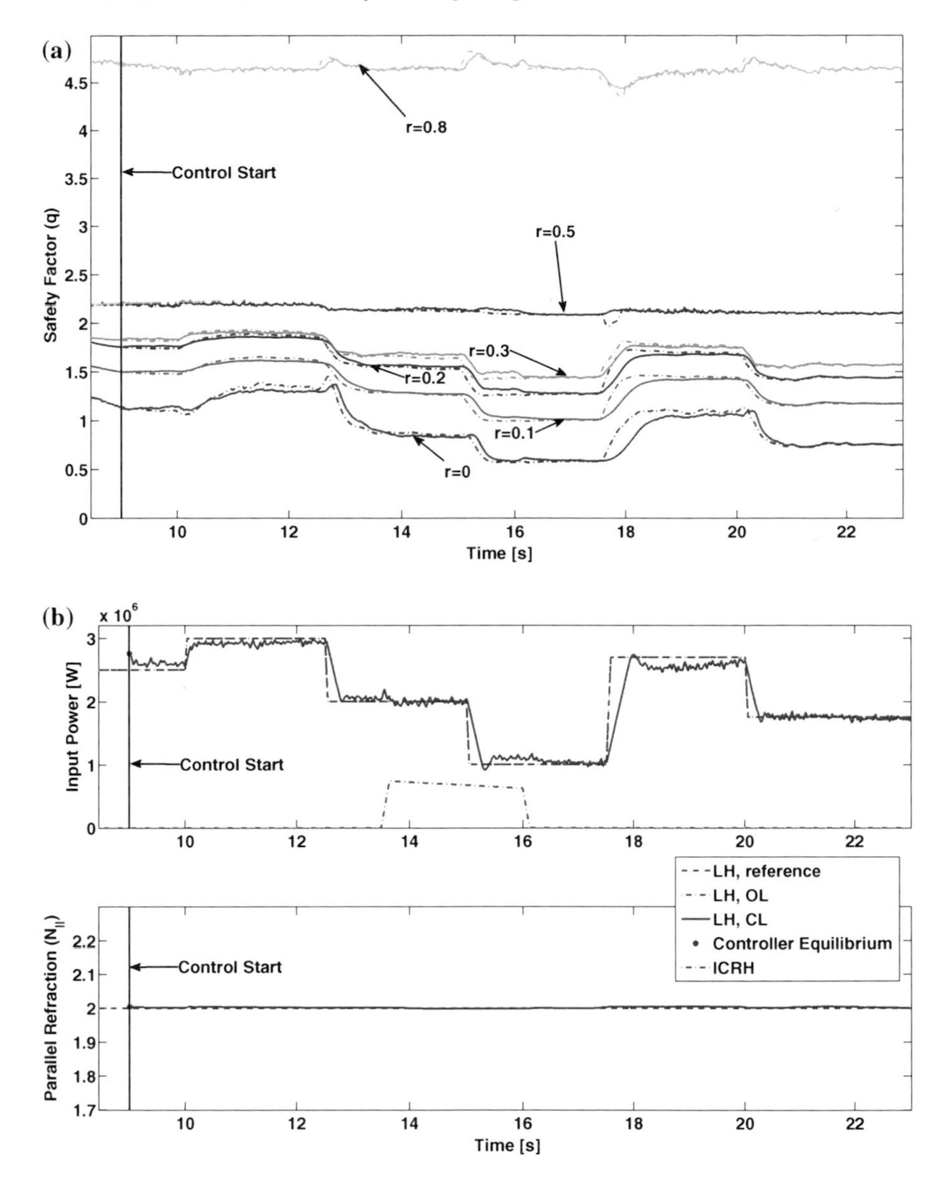

Fig. 5.1 Safety factor profile tracking and radio-frequency antenna control parameter evolution. **a** Tracking of the safety factor profile. *Dashed line* reference q value; *full line* obtained q value. **b** Controller action and ICRH disturbance

the evolution of six points in the safety factor profile. This does not, however, imply that only these six points are being tracked. The beginning of the closed-loop action is marked at 9 s. Solid lines represent the simulated closed-loop values for each variable, while dashed lines represent the reference values (used for the open-loop

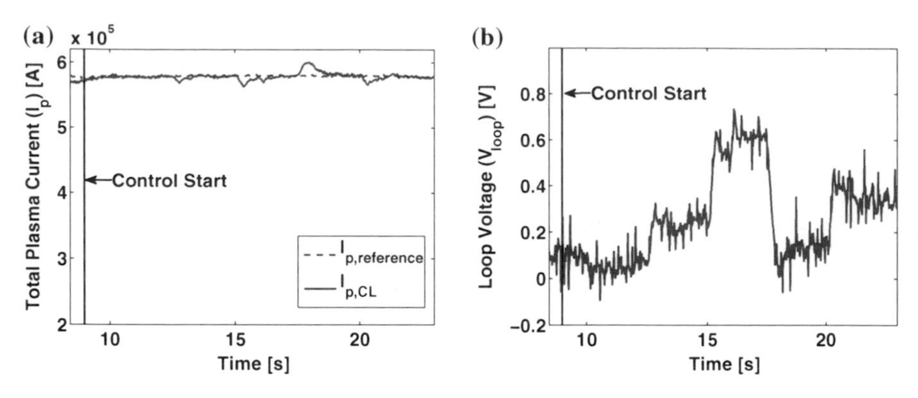

Fig. 5.2 Total plasma current evolution and corresponding loop voltage. **a** Tracking of the total plasma current. **b** Resulting V_{loop} due to I_p tracking

computation of the reference magnetic flux profile). Although the error does not converge to zero when the ICRH disturbance is present (since the target profile is no longer achievable with the available actuation means), the error remains bounded and small. Also, given that the magnitude of the ICRH disturbance is smaller than that used in [7], the magnitude of the error is smaller. When no disturbances are present, the error converges to zero with no evident overshoot. The effect of the boundary condition is more noticeable closer to the plasma edge (for example, at $r = 0.8$). Whenever there are variations of the LH power, the total plasma current is disturbed but eventually returns to its nominal value. Figure 5.1b shows that the controller is able to adequately reconstruct the sequence of engineering parameters used to generate the given trajectory. The engineering parameters corresponding to the equilibrium value given to the controller are marked at the beginning of the control action with a star. The effect of measurement noise on the control action can be appreciated in the small high-frequency variations of P_{lh} which, if desired, can be reduced by decreasing the gain of the controller. Finally, Fig. 5.2a, b show the evolution of the total plasma current and the loop voltage. This figure illustrates that a simple local controller is able to adequately regulate the total plasma current.

5.3.3 Simulation Scenario: METIS, Independent I_p Control, Large Variations of $N_{\|}$, ICRH Heating Disturbance

The second chosen scenario was chosen to test the capacity of the controller to use the other available control parameter. The value of the global variables is the same as in the previous case. The value of P_{lh} is now held constant at 2.7 MW, while the value of $N_{\|}$ is set to vary in the interval [1.8, 2.2]. In this scenario, the differences in shape of the safety factor profile are noticeable, varying the value of $N_{\|}$ tends to change the mid-radius values of the safety factor profile much more than the central

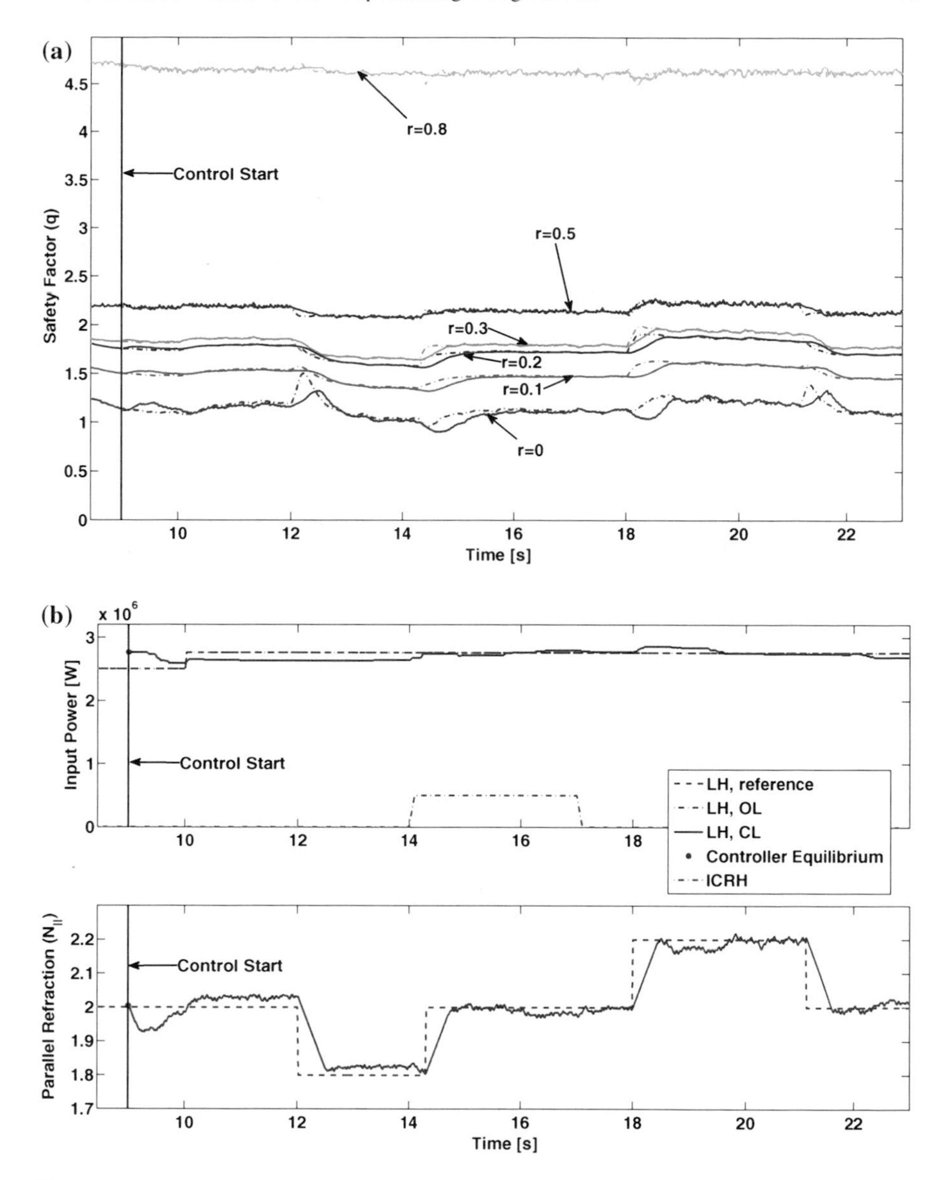

Fig. 5.3 Safety factor profile tracking and radio-frequency antenna parameter evolution. **a** Tracking of the safety factor profile. *Dashed line* reference q value; *full line* obtained q value. **b** Controller action and ICRH disturbance

or edge values. In this case, the power of the ICRH disturbance being less than in the case used in [7], the disturbance is better rejected. At other times, the controller adequately recovers the engineering parameter profiles used to generate the reference (Figs. 5.3 and 5.4).

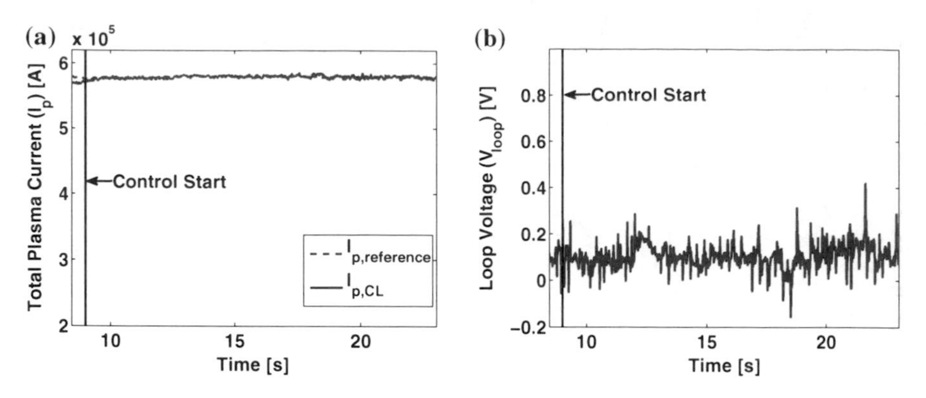

Fig. 5.4 Total plasma current evolution and corresponding loop voltage. **a** Tracking of the total plasma current. **b** Resulting V_{loop} due to I_p tracking

5.4 Some Preliminary Extensions

In this section, we present simulation results illustrating some possible extensions of the methods discussed in this book. All simulation results for TCV, using ECCD (Electron Cyclotron Current Drive) actuation, were computed with the RAPTOR code [9]. Tore Supra results were obtained using the METIS code as discussed in the previous section.

5.4.1 Profile Reconstruction Delays

Based on the sampling time set for the profile reconstruction in [4], the proposed control scheme was tested under simulation adding a 100 ms delay in the control action. The test case is otherwise exactly as described in Sect. 5.3.2. The results are presented in Fig. 5.5. Some tuning was required to avoid overshoots or oscillations and the resulting response is expectedly slower than that shown in Fig. 5.1. However, good convergence is maintained even for references far from the equilibrium value and in the presence of ICRH disturbances. If some overshoots are tolerated, a bigger gain can be used to decrease the settling time. However, since the delay is not negligible compared to the response time obtained in Fig. 5.1, some performance degradation is unavoidable without explicitly addressing the delay (with some predictive structure, for example).

5.4.2 Extension for TCV

In this subsection we present simulation results using ECCD actuation in a simulated shot for the TCV tokamak. A typical shape for the different coefficients and functions

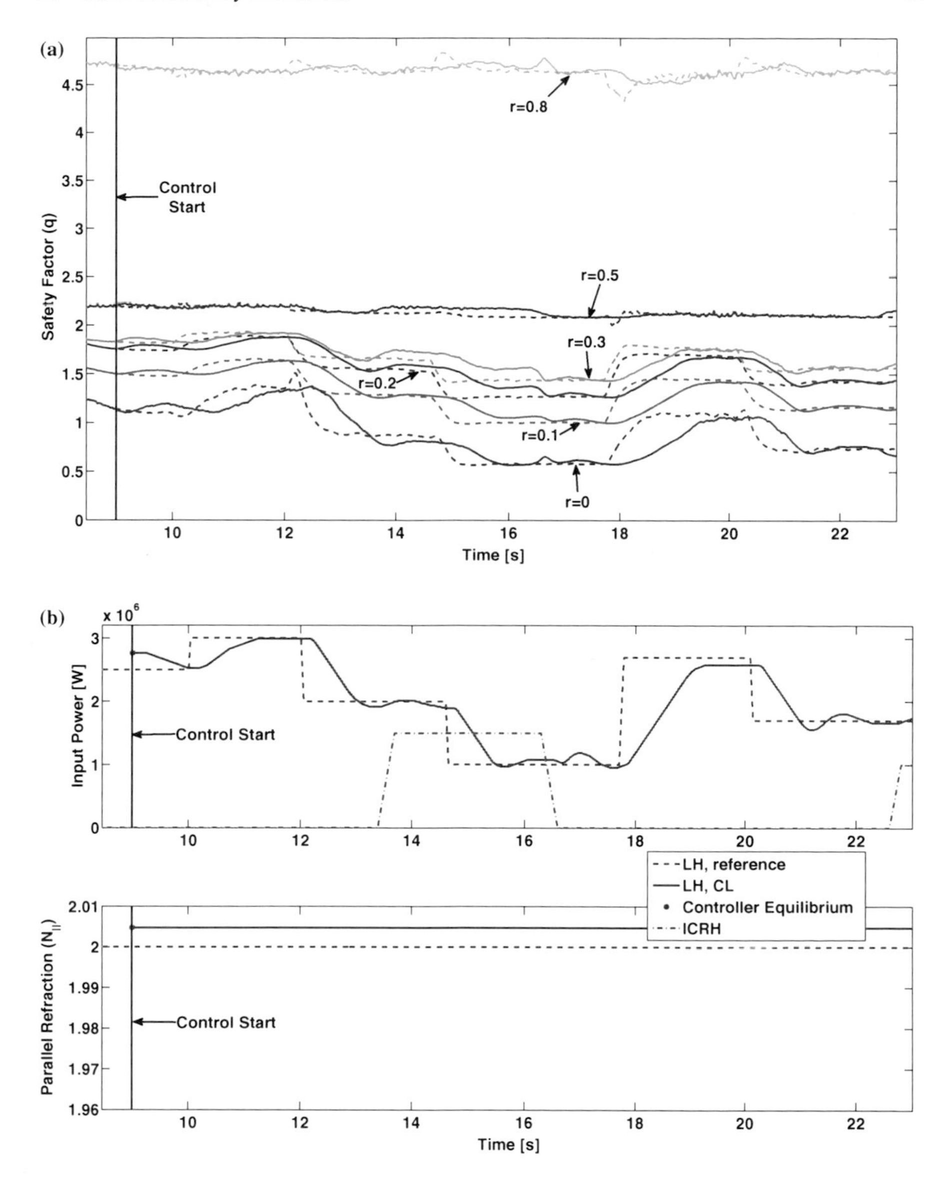

Fig. 5.5 Safety factor profile tracking and radio-frequency antenna control parameter evolution. **a** Tracking of the safety factor profile. *Dashed line* reference q value; *full line* obtained q value. **b** Controller action and ICRH disturbance

in Eq. (2.1) can be found in [9]. These coefficients can be adequately represented by defining the diffusivity coefficients as in (A.1) and slightly modifying the definition of the control u to reflect the small deviations caused by the actual coefficients. Furthermore, the proposed approach is not exclusive to LH actuation. By changing the function u in the optimization problem used to calculate the control action, the same methods previously discussed can be implemented. The reference ECCD model is the one proposed in [9]:

$$j_{ECCD}(\rho, t) = c_{cd} e^{-\rho^2/0.5^2} \frac{T_e}{n_e} e^{-(\rho-\rho_{dep})^2/w_{cd}^2} P_{ECCD}(t) \tag{5.20}$$

where c_{cd} is an experimentally determined coefficient, ρ_{dep} and w_{cd} define the position and width of the current deposit and P_{ECCD} represents the power used by the actuator. All other variables are defined as in Chap. 2. Furthermore, the controller is used around a precomputed open-loop trajectory for the actuators that could be obtained using the methods described in [8] or, possibly, that presented in [12]. For this application, we consider two ECCD actuators with deposits located at $r = 0$ and $r = 0.4$. The two normalized control profiles considered available are shown in Fig. 5.6. The following simulations aim to illustrate the flexibility of the proposed method when considering different non-inductive actuators and a different plasma shape.

The first set of simulations presents the behavior of the undisturbed system in open-loop and closed-loop, see Figs. 5.7 and 5.8, respectively. In both cases the initial power ramp-up is done with an open-loop profile. For the closed-loop simulation, after the ECCD power of the first actuator reaches 2 MW, the controller is activated to accelerate the convergence of the system. Since no disturbances are present, both the open-loop and the closed-loop case reach the desired safety-factor reference at the end of the simulation time. The closed-loop system converges faster than the

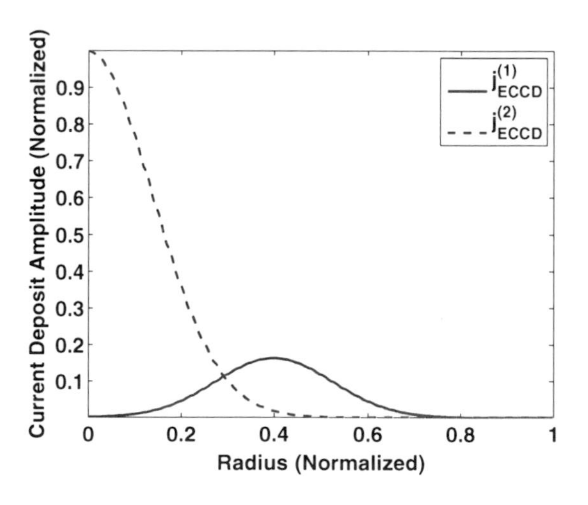

Fig. 5.6 Available ECCD control profiles

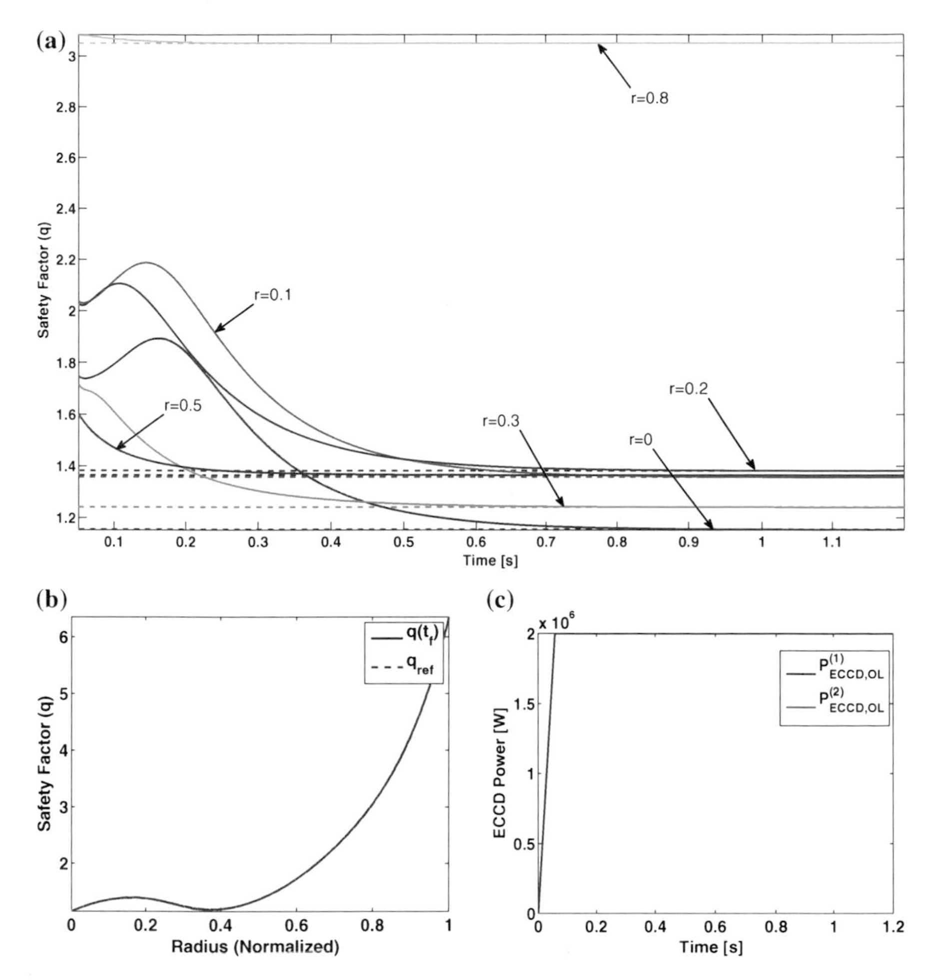

Fig. 5.7 Safety factor profile and open-loop ECCD power evolution. **a** Tracking of the safety factor profile. *Dashed line* reference q value; *full line* obtained q value. **b** Final safety factor profile and reference. **c** Applied ECCD power

open-loop system (~0.5 s versus ~0.8 s), the acceleration being most noticeable for small values of r. Some undershoot is present in the closed-loop behavior, but the control algorithm is tuned so that the safety factor does not reach values under 1. At the end of the simulation time, the values of the ECCD power in both antennas are the same in the open-loop and closed-loop cases.

The second set of simulations includes a disturbance induced by extra ECCD current that peaks at $r = 0.4$. Since one of our available actuators is placed at this particular position, we may expect the controller to be able to greatly attenuate the effect of such a disturbance. The open-loop behavior of the system is shown in Fig. 5.9, while the closed-loop behavior is presented in Fig. 5.10. Figures 5.9

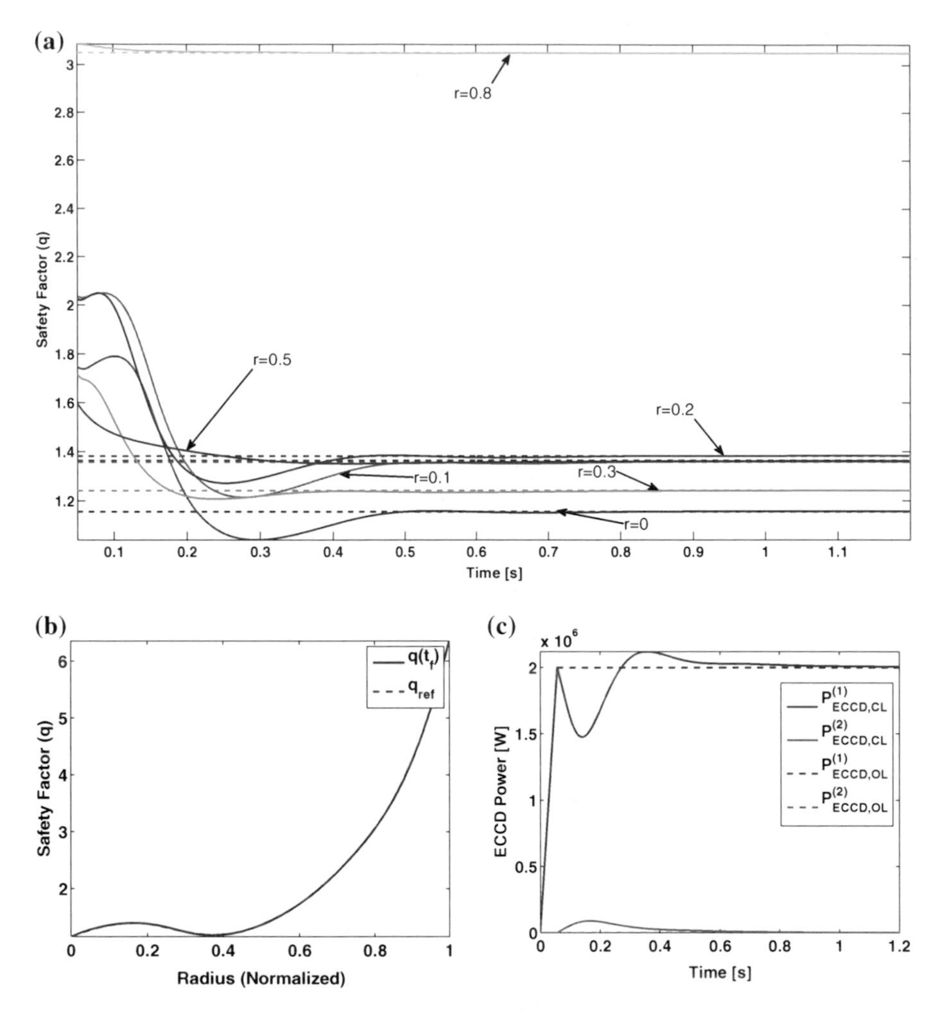

Fig. 5.8 Safety factor profile tracking and closed-loop ECCD power evolution. **a** Tracking of the safety factor profile. *Dashed line* reference q value; *full line* obtained q value. **b** Final safety factor profile and reference. **c** Applied ECCD power and open-loop reference value

and 5.10b illustrate the main interest of the closed-loop action: at the end of the simulation time only the closed-loop system is able to reach the desired q-profile. The final value of the ECCD power in the closed-loop system presents a negative offset that corresponds to the value of the applied disturbance. The power of the central actuator is used only to accelerate the convergence of the system and, at the end of the simulation time, returns to its open-loop value.

The third set of simulations presents a more realistic scenario: a disturbance that cannot be entirely offset by the available actuators will be introduced in the system. In this case, we chose a combination of a heating input (ECRH), located at $r = 0.2$,

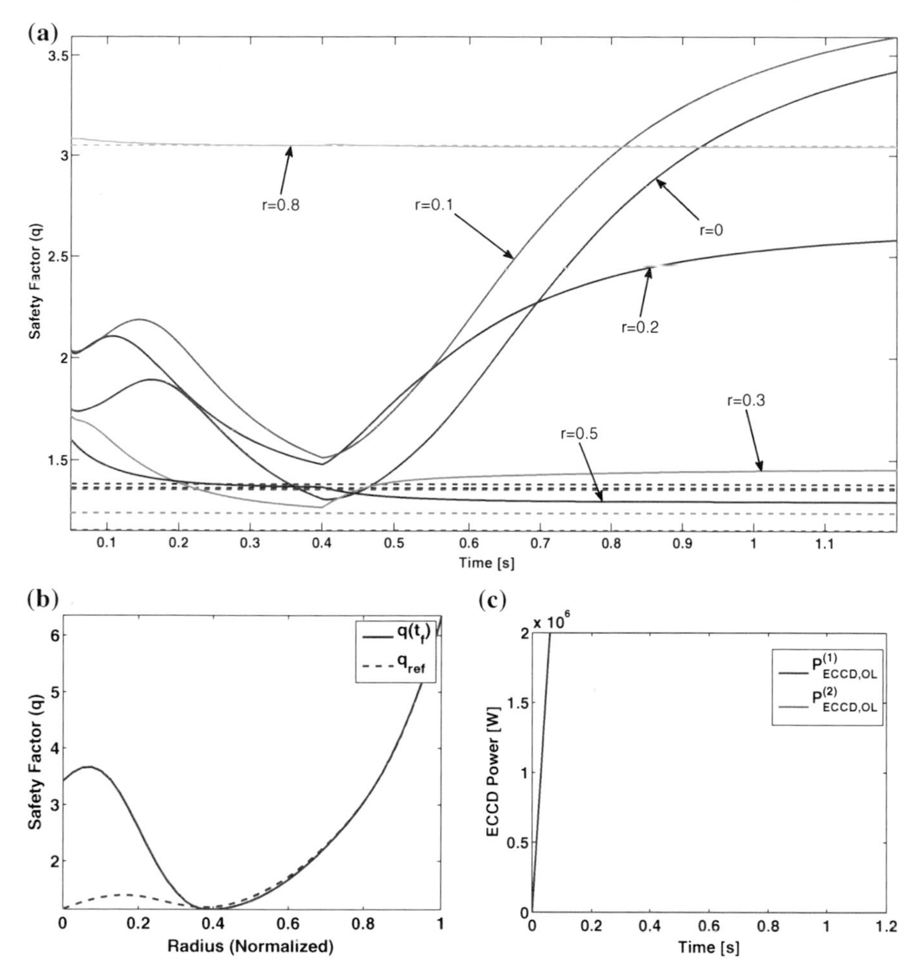

Fig. 5.9 Safety factor profile and open-loop ECCD power evolution with ECCD disturbance applied at $r = 0.4$ for $t \geq 0.4$ s. **a** Tracking of the safety factor profile. *Dashed line* reference q value; *full line* obtained q value. **b** Final safety factor profile and reference. **c** Applied ECCD power

and a current drive input (ECCD), located at $r = 0.4$. The combination of these two inputs cannot be exactly offset by the two available actuators. Nevertheless, comparing the open-loop response (Fig. 5.12) and the closed-loop one (Fig. 5.11), the closed-loop action is clearly beneficial. At the end of the simulation time, the reduction of the error between the reference and the safety factor profile in Fig. 5.11b with respect to Fig. 5.12b is quite apparent.

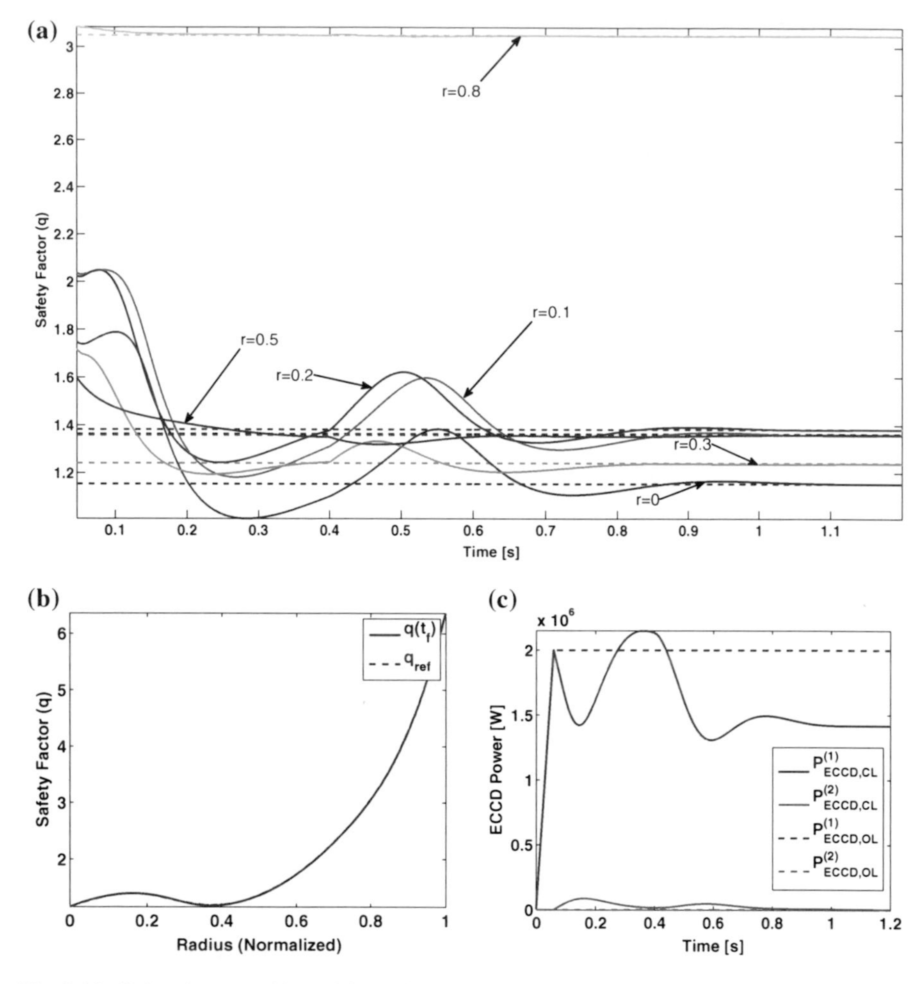

Fig. 5.10 Safety factor profile tracking and closed-loop ECCD power evolution with ECCD disturbance applied at $r = 0.4$ for $t \geq 0.4$ s. **a** Tracking of the safety factor profile. *Dashed line* reference q value; *full line* obtained q value. **b** Final safety factor profile and reference. **c** Applied ECCD power and open-loop reference value

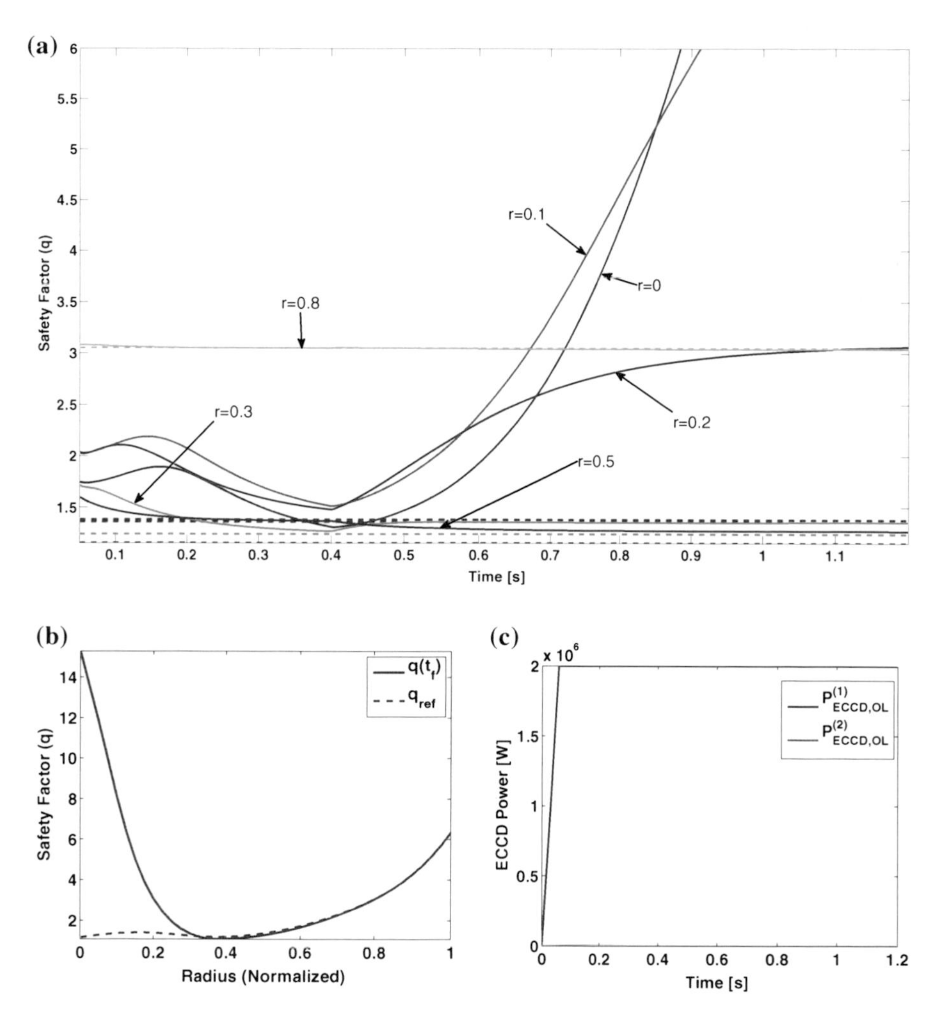

Fig. 5.11 Safety factor profile tracking and closed-loop ECCD power evolution with ECCD and ECRH disturbances applied at $r = 0.4$ and $r = 0.2$, respectively, for $t \geq 0.4$ s. **a** Tracking of the safety factor profile. *Dashed line* reference q value; *full line* obtained q value. **b** Final safety factor profile and reference. **c** Applied ECCD power

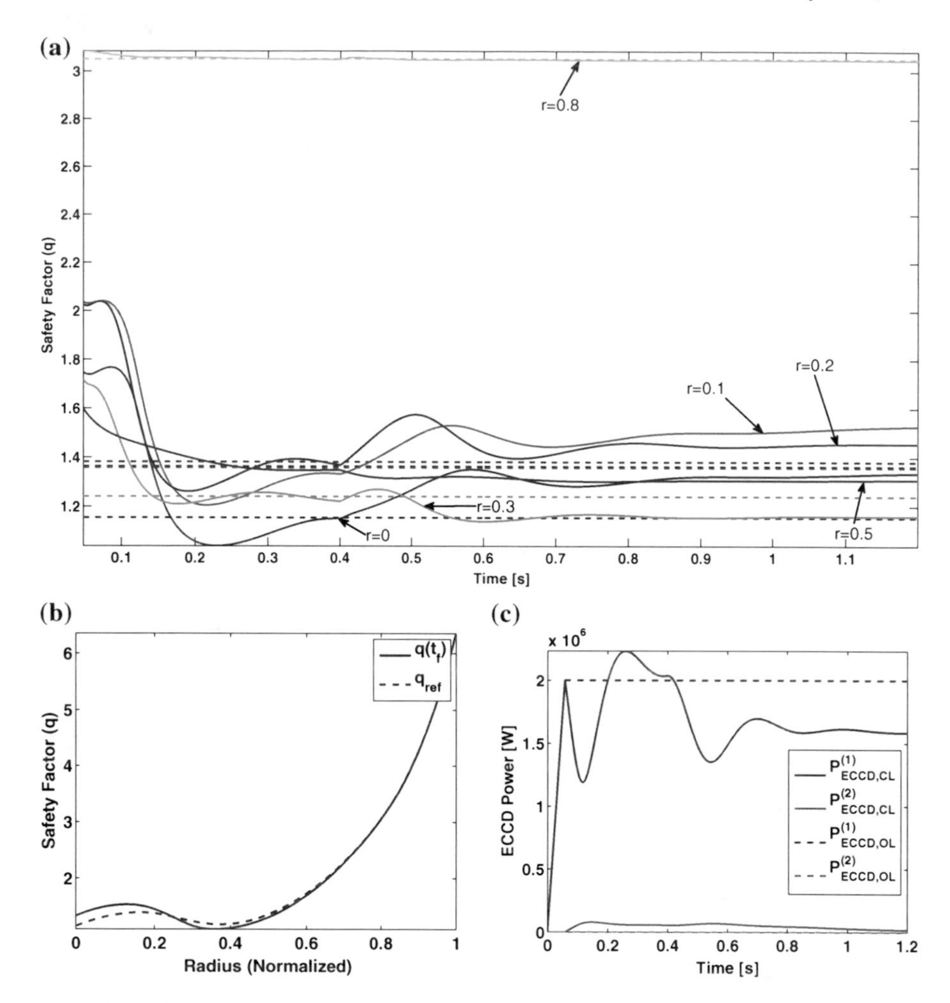

Fig. 5.12 Safety factor profile and open-loop ECCD power evolution with ECCD and ECRH disturbances applied at $r = 0.4$ and $r = 0.2$, respectively, for $t \geq 0.4$ s. **a** Tracking of the safety factor profile. *Dashed line* reference q value; *full line* obtained q value. **b** Final safety factor profile and reference. **c** Applied ECCD power

5.5 Summary and Conclusions

In this chapter, the strict Lyapunov function (4.15) for the poloidal magnetic flux resistive diffusion equation in 1D was modified to consider the couplings that exist between the flux diffusion and the total plasma current circuit (assumed to behave like a transformer, as in [11]). By adding a component representing the energy of the finite-dimensional subsystem, ISS properties similar to those established in the previous chapter can be obtained. A particularly interesting conclusion of this section is that we can always find a controller gain small enough (in terms of the variations of LH power) to guarantee the exponential stability of the interconnected system and that, as we approach a perfectly decoupled system (when the rate of convergence of the finite-dimensional system is much faster than that of the infinite-dimensional one), this constraint becomes less strict.

At this point, most of the difficulties outlined in Chap. 2 have been tackled:

- the time-varying distributed nature of the diffusion coefficients is taken into account (for more details see Appendix A);
- the strong nonlinear shape constraints imposed on the actuators are considered in the control algorithm;
- the robustness of any proposed control scheme with respect to different sources of error and disturbances has been analyzed;
- the coupling between the controller for the infinite-dimensional system and the boundary condition is taken into account;
- the control algorithms considered are implementable in real-time.

Some preliminary extensions of the proposed approach have also been presented. Namely, the addition of a profile-reconstruction delay in the closed-loop simulations and the use of ECCD actuation with TCV parameters.

References

1. J.F. Artaud, METIS user's guide. CEA/IRFM/PHY/NTT-2008.001, 2008
2. J.F. Artaud et al., The CRONOS suite of codes for integrated tokamak modelling. Nucl. Fusion **50**, 043001 (2010)
3. J. Blum, *Numerical Simulation and Optimal Control in Plasma Physics*. Wiley/Gauthier-Villars series in modern applied mathematics. (Wiley, Paris, 1989)
4. J. Blum, C. Boulbe, B. Faugeras, Reconstruction of the equilibrium of the plasma in a Tokamak and identification of the current density profile in real time. J. Comput. Phys. **231**(3), 960–980 (2012)
5. F. Bribiesca Argomedo, C. Prieur, E. Witrant, S. Brémond, *Polytopic control of the magnetic flux profile in a tokamak plasma*. In Proceedings of the 18th IFAC World Congress, (Milan, Italy, 2011), pp. 6686–6691
6. F. Bribiesca Argomedo, E. Witrant, C. Prieur, *D1-Input-to-State stability of a time-varying nonhomogeneous diffusive equation subject to boundary disturbances* (In Proceedings of the American Control Conference, Montréal, Canada, 2012)

7. F. Bribiesca Argomedo, E. Witrant, C. Prieur, S. Brémond, R. Nouailletas, J.F. Artaud, Lyapunov-based distributed control of the safety-factor profile in a tokamak plasma. Nucl. Fusion **53**(3), 033005 (2013)
8. F. Felici, O. Sauter, Non-linear model-based optimization of actuator trajectories for tokamak plasma profile control. Plasma Phys. Control. Fusion **54**, 025002 (2012)
9. F. Felici, O. Sauter, S. Coda, B.P. Duval, T.P. Goodman, J-M. Moret, J.I. Paley, TCV Team, Real-time physics-model-based simulation of the current density profile in tokamak plasmas. Nucl. Fusion **51**, 083052 (2011)
10. M. Goniche et al., *Lower hybrid current drive efficiency on Tore Supra and JET*, 16th Topical Conference on Radio Frequency Power in Plasmas. Park City, 2005
11. F. Kazarian-Vibert et al., Full steady-state operation in Tore Supra. Plasma Phys. Control. Fusion **38**, 2113–2131 (1996)
12. Y. Ou, C. Xu, E. Schuster, T.C. Luce, J.R. Ferron, M.L. Walker, D.A. Humphreys, Design and simulation of extremum-seeking open-loop optimal control of current profile in the DIII-D tokamak. Plasma Phys. Control. Fusion **50**, 115001 (2008)
13. E. Witrant, E. Joffrin, S. Brémond, G. Giruzzi, D. Mazon, O. Barana, P. Moreau, A control-oriented model of the current control profile in tokamak plasma. Plasma Phys. Control. Fusion **49**, 1075–1105 (2007)

Chapter 6
Conclusion

In this book, the problem of controlling the poloidal magnetic flux profile in a tokamak plasma has been studied. The motivation for this problem is the possibility of controlling the safety factor profile necessary to attain and maintain advanced operating modes in a tokamak with enhanced confinement and MHD stability.

The first control approach illustrates a classical method consisting on a spatial discretization of the distributed parameter system before applying established techniques for finite-dimensional systems. Knowing that the the diffusivity coefficients can vary much faster than the flux diffusion time and that some concurrent actuation may be in use (to control other plasma parameters), neglecting the variations of the diffusivity profiles is unsatisfactory. In order to take into account these variations, in Chap. 3, a Polytopic LPV formulation is developed for the discretized system. While it allows us to consider some variations, this approach can be too conservative and fail to provide useful results when a large operating range is considered. Furthermore, the extension of this approach to the gradient of the poloidal flux profile (a more significant physical variable) is far from straightforward and the proposed change of variables (leading to a constant B matrix) cannot be applied.

In order to address the weaknesses discovered in the first proposed approach, in Chap. 4 we present a new approach based on a strict Lyapunov function developed for the open-loop system. Since the physical flux diffusion system is known to be stable, this allows us to construct strongly constrained control laws that guarantee the closed-loop stability of the system while accelerating the rate of convergence and attenuating disturbances. A detailed analysis of robustness properties of control laws based on the constructed ISS-Lyapunov function is developed to illustrate the impact of different design parameters in the controller. This approaches are tested under simulation using both a simple simulation of only the diffusion equation and a more complete model based on [2] that corresponds to the model presented in Chap. 2.

Finally, in Chap. 5, the interconnection of the poloidal magnetic flux diffusion equation with the peripheral dynamics controlling the boundary conditions of the model is explored. An extended Lyapunov function is constructed to show the stabil-

F. Bribiesca Argomedo et al., *Safety Factor Profile Control in a Tokamak*,
SpringerBriefs in Control, Automation and Robotics,
DOI: 10.1007/978-3-319-01958-1_6, © The Author(s) 2014

ity of the interconnected system and the control scheme is tested using METIS. Some extensions are presented, taking into account other important aspects that directly add to the relevance of the proposed approach, mainly the effect of profile-reconstructuon delays and the possibility of extending the results to other tokamaks (done here using the RAPTOR code [1] and TCV parameters) and other non-inductive actuators (in this case ECCD).

The main contributions of this book are:

- the explicit consideration of the time-varying nature of the diffusivity profiles (not limiting their variation to a scalar quantity multiplying a static shape);
- the explicit consideration of the nonlinear dependency of the current deposit on the antenna parameters;
- allowing for actuator saturation;
- the explicit consideration of the couplings existing between the control action and the boundary condition in the stability analysis;
- a thorough characterization of the gains between different sources of error and disturbances and the state;
- the lack of imposed limits on the rate of time variation of the diffusivity profiles to maintain the stability;
- particular provisions for the real-time implementation of the proposed approach for safety factor control on a real tokamak.

The main remaining challenges are:

- extending the stability results to incorporate performance guarantees (to limit overshoots, for instance);
- considering the coupled problem of temperature and current profile control (necessary for burn-control);
- considering the nonlinear impact of the bootstrap current, first to guarantee the stability of the system and then to maximize the bootstrap current fraction;
- finding conditions to guarantee the stability of the system in the presence of delays;
- experimental validation of the proposed approaches.

References

1. F. Felici, O. Sauter, S. Coda, B.P. Duval, T.P. Goodman, J-M. Moret, J.I. Paley, and the TCV Team. Real-time physics-model-based simulation of the current density profile in tokamak plasmas. Nuclear Fusion **51**, 083052 (2011)
2. E. Witrant, E. Joffrin, S. Brémond, G. Giruzzi, D. Mazon, O. Barana, P. Moreau, A control-oriented model of the current control profile in tokamak plasma. Plasma Phys. Control. Fusion **49**, 1075–1105 (2007)

Appendix A
Finding a Lyapunov Function

A.1 Finding a Weighting Function

A.1.1 Considered Set of Diffusivity Profiles

In this Appendix we present a heuristic method to find an adequate weight verifying all the conditions of Theorem 4.1 for a sufficiently large set of resistivity profiles. As mentioned in previous chapters, the stucture of model (2.15) implies that finding a common Lyapunov function for a set of resisitivity profiles guarantees (as in the finite-dimensional case) the stability for any convex combination of these profiles. Unlike the finite-dimensional case, however, the time-derivative of these coefficients does not appear in the evolution of the Lyapunov Function.

Following [1], we select the following set of resistivity profiles:

$$\eta(r,t) = a(t)e^{\int_0^r \phi(\xi,t)d\xi}, \forall (r,t) \in [0,1] \times [0,T) \tag{A.1}$$

with $0 < \underline{a} \leq a(t) \leq \overline{a}$,

$$\phi(r,t) \in \Phi = \left\{\phi(r,t) \in \mathscr{C}^\infty([0,1] \times [0,T]) \mid \forall t \in [0,T],\ \phi(\cdot,t) \in \Lambda\right\}$$

and

$$\Lambda = \left\{\lambda(r) \in \mathscr{C}^\infty([0,1]) \mid \forall r \in [0,1],\ \underline{\lambda} \leq \lambda(r) \leq \overline{\lambda}\right\}$$

This shape is very similar to an exponential but allows for a richer basis for the resistivity profiles (it reduces to an exponential if λ is constant over r).

F. Bribiesca Argomedo et al., *Safety Factor Profile Control in a Tokamak*,
SpringerBriefs in Control, Automation and Robotics,
DOI: 10.1007/978-3-319-01958-1, © The Author(s) 2014

A.1.2 Alternative Sufficient Conditions and Algorithm

Proposition A.1 *With* η *defined as in* (A.1)*, a sufficient condition to apply Theorem 4.1 is the existence of an a.e. twice-differentiable positive function* $f : [0, 1] \to \mathbb{R}^+$ *with piecewise-continuous second derivative such that the following inequality is verified:*

$$f''(r) + f'(r)\left[\lambda(r) - \frac{1}{r}\right] + f(r)\left[\frac{\lambda(r)}{r} - \frac{1}{r^2} + \varepsilon\right] \leq 0 \qquad \text{(A.2)}$$

for every $(r, \lambda) \in (0, 1] \times \Lambda$ *and some positive constant* ε.

See [1] for the proof.

We can rewrite the inequality (A.2) as an equation using a slack variable $w(r, \lambda) \leq 0$. Rewriting the resulting second order differential equation as a system of two first order ODEs we obtain:

$$\begin{bmatrix} f \\ f' \end{bmatrix}' = \begin{bmatrix} 0 & 1 \\ \frac{1}{r^2} - \frac{\lambda(r)}{r} - \varepsilon & \frac{1}{r} - \lambda(r) \end{bmatrix} \begin{bmatrix} f \\ f' \end{bmatrix} + \begin{bmatrix} 0 \\ 1 \end{bmatrix} w(r, \lambda) \qquad \text{(A.3)}$$

We can now introduce the following proposition (see also [1]):

Proposition A.2 *Given an a.e. twice-differentiable positive function with piecewise-continuous second derivative* $f : [0, 1] \to \mathbb{R}^+$*, the following two conditions are equivalent:*

i: *there exists* $w(r, \lambda) \leq 0$ *such that* (A.3) *is verified for all* $(r, \lambda) \in (0, 1] \times \Lambda$*;*
ii: *there exists* $w_2(r) \leq 0$ *such that the following equation is verified for all* $r \in (0, 1]$*:*

$$\begin{bmatrix} f \\ f' \end{bmatrix}' = A(f, r) \begin{bmatrix} f \\ f' \end{bmatrix} + \begin{bmatrix} 0 \\ 1 \end{bmatrix} w_2(r) \qquad \text{(A.4)}$$

where:

$$A(f, r) = \begin{cases} \begin{bmatrix} 0 & 1 \\ \frac{1}{r^2} - \frac{\underline{\lambda}}{r} - \varepsilon & \frac{1}{r} - \underline{\lambda} \end{bmatrix} & \text{if } \frac{f(r)}{r} + f'(r) \leq 0 \\ \begin{bmatrix} 0 & 1 \\ \frac{1}{r^2} - \frac{\overline{\lambda}}{r} - \varepsilon & \frac{1}{r} - \overline{\lambda} \end{bmatrix} & \text{if } \frac{f(r)}{r} + f'(r) > 0 \end{cases}$$

The proof will not be given in this Appendix, however, it is based on the fact that, for a given r, the left-hand side of inequality (A.2) is linear in λ and thus has a maximum value at either $\overline{\lambda}$ or $\underline{\lambda}$. If the inequality holds at this *critical* point, it has to hold for all admissible values of λ.

Remark A.1 Setting the slack variable to zero and solving the inequality as an equation does not provide adequate weighting functions (the solutions have a singularity at zero). For more information on this subject, we refer the reader to [1].

In order to solve the inequality, we propose to set boundary conditions at $r = 1$ and then solve the equation backwards up to $r = 0$ using Algorithm 1 proposed in [1].

Algorithm 1

1: *Set numerical values for the boundary conditions at* $r = 1$, $f(1)$ *and* $f'(1)$, *and for* ε.
2: *Evaluate* $\frac{f(r)}{r} + f'(r)$ *and fix the value of the dynamic matrix* $A(f, r)$ *accordingly, using* (A.4).
3: *Find a numerical solution going backwards until hitting a zero-crossing of* $\frac{f(r)}{r} + f'(r)$, *setting* $w_2(r) = 0$, *and verifying that* $f(r)$ *remains positive. Otherwise, change the boundary conditions or the value of* ε.
4: *Use the values of* $f(r)$ *and* $f'(r)$ *at the zero-crossings of* $\frac{f(r)}{r} + f'(r)$ *as initial values for the next step in solving the equation, switching the dynamic matrix but keeping* $w_2(r) = 0$, *always verifying that* $f(r)$ *remains positive and bounded.*
5: *Repeat 3–4 until either reaching* $r = 0$ *or finding a point such that both elements in the lower row of the A matrix are positive, as well as* f *and* f', *with* $f(r) - rf'(r) > 0$. *If no such point exists before* $r = 0$, *change the boundary conditions or the value of* ε *and start over.*
6: *If* $r = 0$ *has not been reached yet, complete the solution by setting* $w_2(r)$ *to have* $f''(r) = 0$ *for the remaining interval, in order to avoid singularities in the solution near zero.*

Remark A.2 It should be stressed that this algorithm is not guaranteed to provide a solution. For the desired range of λ based on experimental reconstructions in Tore Supra, however, it yields adequate results.

A.2 Numerical Application

A.2.1 Weighting Function

Greatly extending the range used in [1] by using a negative $\underline{\lambda}$, we select the following range of variation for the parameters required in (A.1):

- $\underline{\lambda} = -4$
- $\bar{\lambda} = 7.3$

This set of parameters is larger than the usual set of resistivity profiles encountered in Tore Supra and allows for non-monotonic functions.

The following values were set for the algorithm:

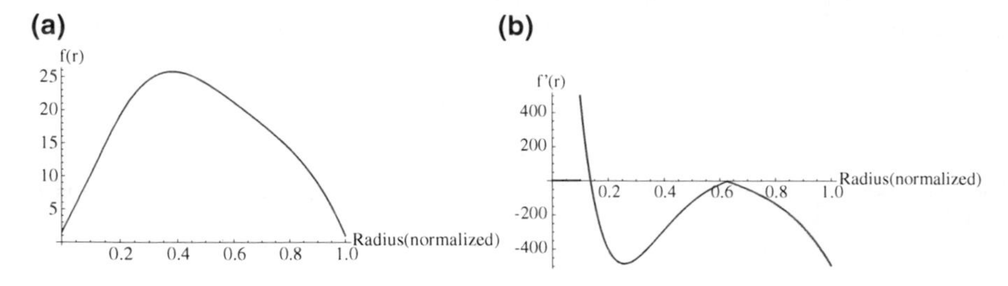

Fig. A.1 Example of weighting function that can be obtained using the presented heuristic. **a** Function f obtained using the heuristic. **b** Piecewise continuous second derivative of function f obtained using the heuristic

- $f(1) = 1$
- $f'(1) = -100$
- $\varepsilon = \pi/2$
- $f''(r) = 0$ for all $r \in [0, 0.1]$

The obtained weight f can be seen in Fig. A.1.

Appendix B
List of Acronyms, Physical Variables and Symbols

List of acronyms

ARE	Algebraic Riccati equation
BIBO	Bounded-input bounded-output
CD	Current drive
ECCD	Electron cyclotron current drive
ICRH	Ion-cyclotron radio heating
ICRH	Ion cyclotron resonance heating
ISS	Input-to-state stability
LHCD	Lower hybrid current drive
LMI	Linear matrix inequality
LPV	Linear parameter-varying
LQR	Linear quadratic regulator
LTI	Linear time-invariant
MHD	Magneto-hydro-dynamics
NBI	Neutral beam injection
PDE	Partial differential equation
RF	Radio frequency
SeDuMi	Self-dual-minimization
TS-35109	Tore-supra shot number 35109
YALMIP	Yet another LMI parser

List of physical variables (unit)

a	Location of the last closed magnetic surface (m)
$B_{\phi 0}$	Toroidal magnetic field at the center (T)
C_2, C_3	Geometric coefficients
c_{cd}	Experimentally determined coefficient
F	Diamagnetic function (Tm)

F. Bribiesca Argomedo et al., *Safety Factor Profile Control in a Tokamak*, SpringerBriefs in Control, Automation and Robotics,
DOI: 10.1007/978-3-319-01958-1,

I_{Ω}	Ohmic current (A)
I_p	Total plasma current (A)
j	Normalized non-inductive effective current density $\mu_0 a^2 R_0 j_{ni}$
j_{bs}	Bootstrap curren (Am^{-2})
j_{eccd}	ECCD current deposit (Am^{-2})
j_{lh}	LHCD current deposit (Am^{-2})
j_{ni}	Non-inductive effective current density (Am^{-2})
$\overline{n}$	Electron average density (m^{-3})
$N_{\parallel}$	Hybrid wave parallel refractive index
$p_{e/i}$	Electron and ion pressure
P_{lh}	Lower hybrid antenna power (W)
q	Safety factor profile $q \doteq d\phi/d\psi$
r	Normalized spatial variable $r \doteq \rho/a$
R_0	Location of the magnetic center (m)
t	Time (s)
$T_{e/i}$	Electron and ion temperature (K)
$\mathcal{V}$	Plasma volume (m^3)
V_{loop}	Toroidal loop voltage (V)
V_{Ω}	Ohmic voltage (V)
w_{cd}	Width of the current deposit (m)
η	Normalized diffusivity coefficient $\eta_{\parallel}/\mu_0 a^2$
$\eta_{\parallel}$	Parallel resistivity (Ωm)
η_{lh}	LH current drive efficiency ($\mathrm{Am}^{-2}\mathrm{W}^{-1}$)
μ_0	Permeability of free space: $4\pi \times 10^{-7}$ (Hm^{-1})
ρ	Equivalent radius of the magnetic surfaces (m)
ρ_{dep}	Position of the current deposit (m)
ϕ	Toroidal magnetic flux profile (Tm^2)
ψ	Poloidal magnetic flux profile (Tm^2)

List of symbols

$f_{r,\,rr}$	First and second derivative of a given function f
L^2, $\|\cdot\|_2$	The set of square-integrable functions equipped with the norm $\|\cdot\|_2$
$M \succ 0$	The matrix M is positive definite
V	Lyapunov function

Reference

1. F. Bribiesca Argomedo, E. Witrant, C. Prieur, D1-Input-to-State stability of a time-varying nonhomogeneous diffusive equation subject to boundary disturbances, in *Proceedings of the American Control Conference*, Montréal, Canada (2012), pp. 2978–2983

Index

F. Bribiesca Argomedo et al., *Safety Factor Profile Control in a Tokamak*,
SpringerBriefs in Control, Automation and Robotics,
DOI: 10.1007/978-3-319-01958-1, © The Author(s) 2014

Printed by Publishers' Graphics LLC
DBT140121.15.18.224